Model Predictive Control - Theory and Applications

Edited by Constantin Volosencu

Published in London, United Kingdom

Model Predictive Control - Theory and Applications
http://dx.doi.org/10.5772/intechopen.1001523
Edited by Constantin Volosencu

Contributors
Balachandar Guduri, Goksel Kizir, Abdul Furkan Kanburoglu, Hiroshi Hanaizumi, Song Ho Kim, Kwang Rim Song, Nesrine Ben Afia, Hassani Messaoud, Francis Assadian, George Martin, K. Karthigaiselvan, Rames Chandra Panda

First published in London, United Kingdom, 2023 by IntechOpen
IntechOpen is the global imprint of INTECHOPEN LIMITED, registered in England and Wales, registration number: 11086078, 5 Princes Gate Court, London, SW7 2QJ, United Kingdom

British Library Cataloguing-in-Publication Data
A catalogue record for this book is available from the British Library

Additional hard and PDF copies can be obtained from orders@intechopen.com

Model Predictive Control - Theory and Applications
Edited by Constantin Volosencu
p. cm.
Print ISBN 978-1-80355-988-9
Online ISBN 978-1-80355-987-2
eBook (PDF) ISBN 978-1-80355-989-6

Meet the editor

Prof. Dr. Constantin Voloşencu graduated as an engineer from Politehnica University Timişoara, Romania, where he also obtained a doctorate degree. He is currently a full professor in the Department of Automation and Applied Informatics at the same university. Professor Voloşencu is the author of 10 books, 7 book chapters, and more than 160 papers published in journals and conference proceedings. He has also edited thirteen books and has twenty-seven patents to his name. He is a manager of research grants, editor-in-chief and member of international journal editorial boards, a former plenary speaker, a member of scientific committees, and chair at international conferences. His research is in the fields of control systems, control of electric drives, fuzzy control systems, neural network applications, fault detection and diagnosis, sensor network applications, monitoring of distributed parameter systems, and power ultrasound applications. He has developed automation equipment for machine tools, spooling machines, high-power ultrasound processes, and more.

Contents

Preface

Model predictive control (MPC) is an advanced method of process control based on an optimal control technique in which the calculated control actions minimize a cost function for a constrained dynamical system over a finite, receding horizon. At each time step, a model predictive controller receives or estimates the current state of the plant. MPC can anticipate future events and take control actions accordingly. Model predictive controllers rely on linear mathematical models of the process, obtained by system identification. MPC is implemented as a discrete time control system, using digital equipment. At first, in the 1980s, it was used in the control of chemical processes. Then, it began to be used in power systems and power electronics. Now, MPC is a powerful method for engineers and scientists that has applications in many technological domains, assuring the increased efficiency of complex control systems.

This book's subject matter ranges from new theoretical control structures to control applications in the chemical industry and electrical and mechanical engineering. It promotes MPC theory in some practical applications of these engineering domains. The book covers some theoretical concepts of MPC and includes new methodologies, control structures, and applications. The applications in this project, through a high technical level of scientific achievements, improve the properties and performances of complex process control systems. Published studies demonstrate the capability of MPC in various practical fields, enhancing human understanding and improving the specialist's ability to analyze, design, and implement high-performance algorithms. Applications are presented that focus on the methodologies used, with case studies, implementation, measurement, and testing issues. These applications involve domains such as the chemical industry (e.g., atmospheric plasma spray process, reactive separation technique in the natural gas sweetening process, or thermal treatments of raw materials) and motion control (e.g., control of permanent magnet synchronous motor, monitoring of pose of a walking person, or lateral control of a passenger vehicle). New examples of complex control systems for technological development platforms are promoted. This book is designed for specialists who are interested in applications and solutions based on MPC in various fields.

The book includes seven chapters. Chapter 1 presents a solution based on model reference adaptive control to improve the transient performance of a closed loop system under uncertainties and external disturbance, with application in the atmospheric plasma spray process. Chapter 2 presents an application of MPC and a PID controller for position regulation of a permanent magnet synchronous motor with practical experimentation. Chapter 3 presents a method for stereo measurement of 3D poses of a walking person based on a puppet model and a low-rank filter. Chapter 4 presents a perspective study to design a hierarchically structured MPC as the online set-point optimizer for maximizing economic efficiency over a long time horizon to meet product quality requirements. The chapter proposes a new method of using cooperative actions of multi-agent Q-learning for determining the set-point of controllers to approximate the temperature trajectory of the raw material through the cascaded three-step kiln process to an integrated reference trajectory. Chapter 5 presents a theoretical study on the control of a particular class of discreet event systems, a subclass of conflicted timed event graphs with the use of dioid algebra. Chapter 6 proposes a short theoretical comparative study on some methodologies for lateral control of a passenger vehicle. Finally, Chapter 7 presents a theoretical introduction to possible applications of MPC in reactive separation techniques, particularly in the natural gas sweetening process.

The editor thanks the authors for their excellent contributions. The editor also wishes to thank Author Service Manager Ivana Barac and the staff at IntechOpen for their assistance throughout the publication process.

Constantin Voloşencu
Department of Automation and Applied Informatics,
"Politehnica" University Timişoara,
Timişoara, România

Chapter 1

Modified Robust Adaptive Process Control with Improved Transient Performance and Its Application to Atmospheric Plasma Spray Process

Balachandar Guduri

Abstract

This chapter presents the σ-modification and the low-frequency learning to the model reference adaptive control (MRAC) to improve the transient performance of a closed-loop control system under uncertainties and external disturbances. The σ-modification improved the robustness to external bounded disturbances, and the low-frequency learning enabled fast adaption without creating high-frequency oscillations. The feasibility of the resulting robust adaptive control architecture is tested for the atmospheric plasma spray process (APSP) to achieve consistency in ceramic coatings through numerical simulations. The robustness and adaptability of the modified and the standard MRAC architecture are investigated under artificially induced disturbances. The proposed architecture shows better performance than the standard MRAC under external disturbances. The results depict the improved transient performance with faster convergence and smooth variations in closed-loop control signals without creating high-frequency oscillations or steady-state errors.

Keywords: plasma spray process, robust adaptive control, ceramic coating, model reference adaptive control, transient performance

1. Introduction

Adaptive control is a popular control methodology for various applications due to its unique ability to accommodate system parametric and structural uncertainties caused by payload variations, component failures, and external disturbances [1]. The model reference adaptive control (MRAC) is a well-known adaptive control approach. The controller gains of the MRAC are updated based on the tracking error between the measured outputs of the uncertain system with outputs of the selected reference model such that these gains force the uncertain system to track the reference model adaptively. In many applications, the MRAC performs satisfactorily and achieves asymptotic system performance without excessive reliance on the system models [2]. However, it suffers the lack of robustness properties in the presence of model uncertainty and external disturbances, which result in various instabilities, such as parameter drifts, high-gain instability, instability resulting from fast adaption, and high-frequency instability [3–5].

For improving the robustness and transient performance and avoiding high-frequency instability in the presence of uncertainties and external disturbances, a significant research effort has been carried out by modifying the adaptive laws of the MRAC and its architecture [6]. First, to improve the robustness, many researchers have proposed various robust MRAC (we referred to as R-MRAC) schemes by modifying the adaptive laws of the MRAC. The R-MRAC using fixed modification [5] achieves the desired robustness and performance without explicit knowledge of plant dynamics and bounds of the external disturbances. The modified MRAC achieves the boundedness of the closed-loop signals; however, the asymptotic convergence of the tracking error in the absence of the disturbances is not achievable. Apart from this, the MRAC and the R-MRAC using modification schemes can create high-frequency oscillations in control responses when the fast adaption using the high adaptive gains is enabled; thus, it leads to process instability. Additionally, the fixed modification can create a steady-state error and often may lead to a bursting phenomenon [5]. Yucelen and Haddad have proposed low-frequency learning to the MRAC to avoid high-frequency oscillation in control responses in the presence of high adaptive gains [2]. In their work, it was implemented to a nonlinear uncertain plant dynamic without a disturbance term. The fast adaption is achieved using high-adaptive gains by filtering out the high-frequency content in the control responses. This scheme preserves the ideal property of the MRAC, that is, the asymptotic convergence of the tracking error to zero in the absence of the disturbance term.

The motivation, then, is to develop a robust adaptive process control scheme with an improved transient performance by overcoming the issues related to the MRAC. This chapter presents a modified robust model reference adaptive control (we call it MR-MRAC) architecture by incorporating two modifications to the MRAC. First, to achieve robustness in the presence of bounded disturbances, the R-MRAC is considered by incorporating σ-modification to the MRAC. Next, to avoid steady-state error associated with σ-modification and to enable fast adaption using high-adaptive gain, we used low-frequency learning with low-pass filters for the gain estimations. The resulting MR-MRAC scheme achieves the boundedness of all responses of closed-loop control and convergence of the tracking error to a small bound in the presence of bounded disturbances. The feasibility of the MR-MRAC architecture is tested for the atmospheric plasma spray process (APSP), and its adaptability and robustness in the presence of external disturbances to achieve desired consistency in the outputs have been investigated.

The APSP has become one of the most efficient and reliable techniques to produce a wide variety of coatings, such as low porosity, thermal barrier, wear, and corrosive-resistant coatings, in different applications. The APSP is a common technique to produce ceramic coatings and is capable of producing functionally graded coatings (FGCs) [7]. Due to the demand for high-volume manufacturing of these coatings, reproducibility and repeatability of the coating quality are of prime importance.

As depicted in **Figure 1**, the APSP includes plasma generation, plasma and in-flight particle interaction, and deposit formation on a substrate. The plasma produced by the injection of Ar and H_2 into the torch exits from it at a very high temperature and speed. Powder particles injected into the plasma travel with it to the substrate and, upon impact, are deposited on it. The y- and the z-axes are, respectively, along and perpendicular to the plasma jet. Particles' axial velocity and temperature are monitored in the 1-cm wide observation window located near the substrate to be coated.

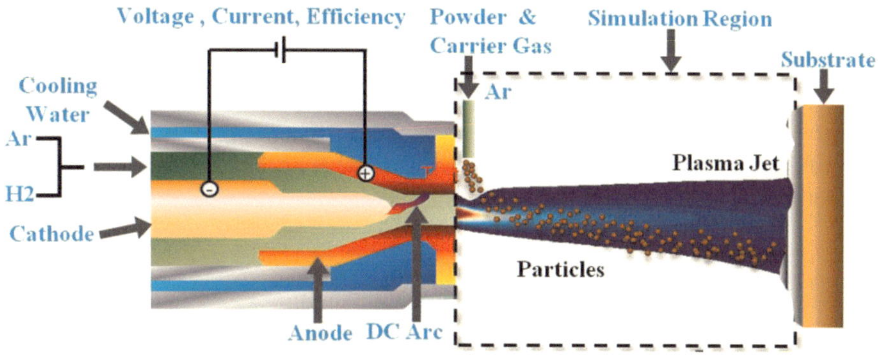

Figure 1.
Schematics of the APSP for producing ceramic coating using a single torch—Single injector system [8].

Complex interactions between plasma and particles significantly [9, 10], vary coating properties that affect the process repeatability [11]. Due to large velocity and temperature gradients in a plasma jet, even small changes in process parameters can significantly alter mean particles' states (mean values of temperature, and axial velocity at the instant of striking the substrate), and thus the quality of the coatings [12]. Even with the operating parameters set constant during the process, the particles' states before impacting the substrate can change over time due to noise variables such as nozzle wear, injector wear, pulsing of powder due to leaks, worn parts in the powder feeder, and powder dampness.

We presented the complete details of the design of the adaptive process control for APSP (validation of numerical simulations, screening of process parameters, finding optimal input parameters, system identification, and the control design) are given in Refs. [13] and [14] for ceramic coating using ZrO_2 powder with a single torch – single injector system and in Ref. [15] for FGCs using a mixture of the NiCrAlY and the ZrO_2 particles with a single torch – two injector system. This chapter discusses the proposed MR-MRAC architecture for generating ceramic coating using ZrO_2 powder. The control objective is to maintain the mean axial velocity and temperature of the particles collected in the observation window as the desired values within small bounds by attenuating the influence of external disturbances.

The remainder of this chapter is organized as follows. Section 2 presents preliminaries about the standard MRAC. Section 3 provides the proposed modified robust model reference adaptive control (MR-MRAC) and its stability properties for multi-inputs and multi-outputs (MIMO) linear time-invariant system with bounded external disturbances. The adaptive and robust performance of the proposed controller using numerical simulations for illustrative examples related to the APSP is presented in Section 4. Finally, the conclusions are summarized in Section 5.

2. MRAC problem formulation

We choose the following plant dynamics with m inputs and n outputs and inclusion of disturbance term:

$$\dot{y}(t) = Ay(t) + Bu(t) + d(t), y(0) = y_0 \tag{1}$$

where $y(t) \in \mathbb{R}^n$ is the output vector, $u(t) \in \mathbb{R}^m$ is the control input vector, $A \in \mathbb{R}^{n\times n}$ and $B \in \mathbb{R}^{n\times m}$ are unknown constants and A is Herwitz, $d(t)$ is the vector of unknown bounded smooth disturbances, which satisfies $\|d(t)\|_2 \le d_{max}$, $\left\|\dot{d}(t)\right\|_2 \le \dot{d}_{max}$ with unknowns and positive bounds $d_{max}(\ge 0), \dot{d}_{max}(\ge 0)$. The pair (A, B) is assumed to be controllable, $y(t)$ is controllable, and $u(t)$ is measurable and locally bounded.

The reference model is described as follows:

$$\dot{y}_m(t) = A_m y_m(t) + B_m r(t), y(0) = y_{m0} \tag{2}$$

where $y_m(t) \in \mathbb{R}^n$ is the reference model output vector, $r(t) \in \mathbb{R}^m$ is a bounded piecewise continuous reference input, $A_m(t) \in \mathbb{R}^{n\times n}$ is Hurwitz and $B_m(t) \in \mathbb{R}^{n\times m}$. The reference model and reference input $r(t)$ are chosen such that $y_m(t)$ represents a desired output trajectory that system output $y(t)$ must follow.

The objective is to design a control input $u(t)$ in Eq. (1) such that the closed loop system has bounded control signals and $y(t)$ adaptively tracks $y_m(t)$ with bounded errors in the presence of uncertainties. We assume there exist ideal gains $K^* \in \mathbb{R}^{m\times n}$ and $L^* \in \mathbb{R}^{m\times m}$ are chosen to satisfy the algebraic equations

$$A - BK^* = A_m, \;\; BL^* = B_m \tag{3}$$

If the matrices A and B are known, use the following controller:

$$u(t) = -K^* y(t) + L^* r(t) \tag{4}$$

In the absence of disturbance guarantees that $y(t) = y_m(t), \forall t \ge 0$ when $y(0) = y_m(0)$ or tracking error $|y(t) - y_m(t)| \to 0$ exponentially when $y(0) \ne y_m(0)$ for any bounded $r(t)$. Since the matrices A *and* B are unknown, the ideal gain matrices K^* *and* L^* are also unknown. Instead of using the ideal control law listed in Eq. (4), the following adaptive control law given by Eq. (5) is considered:

$$u(t) = -K(t)y(t) + L(t)r(t) \tag{5}$$

where $K(t) \in \mathbb{R}^{m\times n}$ and $L(t) \in \mathbb{R}^{m\times m}$ are real-valued gain matrices and are the estimates of the ideal gains at time t. The following adaptive law of the MRAC scheme is used to calculate gains [5]:

$$\begin{aligned} \dot{K}(t) &= \Lambda B_m^T \tilde{P} e(t) y^T(t)\, sgn\,(l), \quad & K(0) = K_0 \\ \dot{L}(t) &= -\Lambda B_m^T \tilde{P} e(t) r^T(t)\, sgn\,(l), \quad & L(0) = L_0 \end{aligned} \tag{6}$$

Here, $\Lambda = \Lambda^T$ and $\tilde{P} = \tilde{P}^T$ are $m \times m$ and $n \times n$ positive definite matrices, respectively. The matrix P is the solution of the Lyapunov equation:

$$A_m^T \tilde{P} + \tilde{P} A_m + Q = 0 \tag{7}$$

For any $Q = Q^T > 0$. For either positive or negative definite L^*, there exists a constant adaptive rate matrix $\Lambda \in \mathbb{R}^{m\times m}$ such that $\Lambda^{-1} = L^*\, sgn\,(l)$, where $l = 1$ if L^* is positive definite and $l = -1$ if L^* is negative definite. The dynamics of the tracking error, $e(t) \equiv y(t) - y_m(t)$ can be written as:

$$\dot{e}(t) = A_m e(t) + B_m {L^*}^{-1}\left(-\tilde{K}(t)y(t) + \tilde{L}(t)r(t)\right) + d(t), e(0) = e_0 \tag{8}$$

where $\tilde{K}(t) \triangleq K(t) - K^*$ and $\tilde{L}(t) \triangleq L(t) - L^*$ are parameter error.

Since $r(t)$ is bounded, this MRAC scheme can guarantee that the closed-loop signals are bounded and outputs of the reference model $y_m(t)$ asymptotically converge to the outputs $y(t)$ in the absence of external disturbance, that is, $d(t) = 0$. If the adaption rate is increased to achieve faster convergence, the transient behavior of $y(t)$ and $u(t)$ cannot be guaranteed and lead to high-frequency oscillations in the closed-loop control signals [6].

3. Modified robust model reference adaptive control (MR-MRAC) formulation

The foregoing MRAC scheme may suffer from instabilities, such as parameter drift, high gains, and/or fast adaption [16]. Some of these could be avoided using the robust MRAC (R-MRAC) by modifying Eqs. (6)–(9) as proposed by Ioannou and Kokotovic [17].

$$\begin{aligned} \dot{K}(t) &= \Lambda\left(B_m^T \tilde{P} e(t) y^T(t)\, sgn\,(l) - \sigma K(t)\right), \quad K(0) = K_0 \\ \dot{L}(t) &= \Lambda\left(-B_m^T \tilde{P} e(t) r^T(t)\, sgn\,(l) - \sigma L(t)\right), \quad L(0) = L_0 \end{aligned} \tag{9}$$

To prevent the steady-state error due to the σ-modification and high-frequency oscillations in MRAC, we are using low-frequency learning with low-pass filters. We consider low-pass filters $K_f(t) \in \mathbb{R}^{m \times n}$, $L_f(t) \in \mathbb{R}^{m \times m}$ presented in Ref. [2], which are weight estimates of gains $K(t), L(t)$, respectively, and are given by

$$\begin{aligned} \dot{K}_f(t) &= \lambda\left(K(t) - K_f(t)\right), \quad K_f(0) = K_0 \\ \dot{L}_f(t) &= \lambda\left(L(t) - L_f(t)\right), \quad L_f(0) = L_0 \end{aligned} \tag{10}$$

where $\lambda > 0$ is a design parameter that serves as the cutoff frequency to suppress the high-frequency oscillations in the closed-loop control system. These low-pass filters only pass gains with frequencies lower than that of cutoff values and the rest will be attenuated. Low-frequency learning is incorporated in σ-modified adaptive laws presented in Eq. (9) to enforce a distance criterion between the estimated gains $K(t), L(t)$ and estimated filter gains $K_f(t), L_f(t)$. This results in a minimization problem containing the following error criteria:

$$\begin{aligned} \min_{K_f} J_K\left(K, K_f\right) &= \frac{1}{2}\left\|K(t) - K_f(t)\right\|_F^2 \\ \min_{L_f} J_L\left(L, L_f\right) &= \frac{1}{2}\left\|L(t) - L_f(t)\right\|_F^2 \end{aligned} \tag{11}$$

The negative gradient of the cost function with respect to gain direction leads to

$$-\frac{\partial J_K}{\partial K} = -\left(K(t) - K_f(t)\right); \; -\frac{\partial J_L}{\partial L} = -\left(L(t) - L_f(t)\right) \tag{12}$$

The implementation of the MR-MRAC to the APSP is the same as that of Refs. [14] and [18] except the following form of the adaptive law is used:

$$\begin{aligned} \dot{K}(t) &= \Lambda\left(B_m^T\tilde{P}ey^T - \sigma(K - K_f)\right), \quad K(0) = K_0 \\ \dot{L}(t) &= \Lambda\left(-B_m^T\tilde{P}er^T - \sigma(L - L_f)\right), \quad L(0) = L_0 \end{aligned} \tag{13}$$

Define $e(t)\triangleq y(t) - y_m(t)$, $\tilde{K}(t)\triangleq K(t) - K^*$, $\tilde{L}(t)\triangleq L(t) - L^*$, $\tilde{K}_f(t) = K_f(t) - K^*$, and $\tilde{L}_f(t) = L_f(t) - L^*$ are the parametric errors. The resulting dynamics for the parametric errors based on the σ-modification of MRAC's adaptive laws with low-pass filter estimation are given as:

$$\begin{aligned} \dot{e}(t) &= A_m e + B_m L^{*\,-1}\left(-\tilde{K}y + \tilde{L}r\right) + d(t), e(0) = e_0 \\ \dot{\tilde{K}}(t) &= \Lambda\left(B_m^T\tilde{P}ey^T - \sigma(\tilde{K} - \tilde{K}_f)\right), \tilde{K}(0) = \tilde{K}_0 \\ \dot{\tilde{L}}(t) &= \Lambda\left(-B_m^T\tilde{P}er^T - \sigma(\tilde{L} - \tilde{L}_f)\right), \tilde{L}(0) = \tilde{L}_0 \\ \dot{\tilde{K}}_f(t) &= \lambda\left(\tilde{K} - \tilde{K}_f\right), \tilde{K}_f(0) = \tilde{K}_0 \\ \dot{\tilde{L}}_f(t) &= \lambda\left(\tilde{L} - \tilde{L}_f\right), \tilde{L}_f(0) = \tilde{L}_0 \end{aligned} \tag{14}$$

Theorem: *Consider the plant dynamics shown in* Eq. (1), *the reference model in* Eq. (2), *the control law in* Eq. (5). *Then, the solution* $\left(e(t), \tilde{K}(t), \tilde{L}(t), \tilde{K}_f(t), \tilde{L}_f(t)\right)$ *of the dynamical system given by* Eq. (14) *is uniformly bounded for all* $\left(e_0, \tilde{K}_0, \tilde{L}_0, \tilde{K}_{f0}, \tilde{L}_{f0}\right) \in D_\alpha$ *where* D_α *is a compact positive invariant set, with ultimate bound.*

$$\|e(t)\|_2 < 2\frac{\lambda_{max}(\tilde{P})d_{max}}{\lambda_{min}(Q)}, t \geq 0 \tag{15}$$

Additionally, the L_2 – norm of error is bounded from above for $t \geq 0$ as indicated below:

$$\|e\|_{L_2}^2 \leq \frac{1}{\left|\lambda_{min}(Q) - \lambda_{max}(\tilde{P})\right|}\left(V_0 + \lambda_{max}(\tilde{P})d_{max}^2\right). \tag{16}$$

Proof: We consider the following Lyapunov candidate function:

$$V\left(e, \tilde{K}, \tilde{L}, \tilde{K}_f, \tilde{L}_f\right) = e^T\tilde{P}e + tr\left(\tilde{K}^T\Gamma\Lambda^{-1}\tilde{K}\right) + tr\left(\tilde{L}^T\Gamma\Lambda^{-1}\tilde{L}\right) + \frac{\sigma}{\lambda}tr\left(\tilde{K}_f^T\Gamma\tilde{K}_f\right) + \frac{\sigma}{\lambda}tr\left(\tilde{L}_f^T\Gamma\tilde{L}_f\right) \tag{17}$$

where $\tilde{P} = \tilde{P}^T > 0$ satisfies the Lyapunov equation in Eq. (7). Since $\tilde{P}, \Gamma, \Lambda$ are positive definite, $\sigma > 0$, and $\lambda > 0$; $V(0, 0, 0, 0, 0) = 0$ and $V\left(e, \tilde{K}, \tilde{L}, \tilde{K}_f, \tilde{L}_f\right) > 0$ for all $\left(e, \tilde{K}, \tilde{L}, \tilde{K}_f, \tilde{L}_f\right) \neq (0, 0, 0, 0, 0)$. Also, $V\left(e, \tilde{K}, \tilde{L}, \tilde{K}_f, \tilde{L}_f\right)$ is radially unbounded. The time derivative along system trajectories is

$$\dot{V} = \dot{e}^T\tilde{P}e + e^T\tilde{P}\dot{e} + 2tr\left(\tilde{K}^T\Gamma\Lambda^{-1}\dot{\tilde{K}}\right) + 2tr\left(\tilde{L}^T\Gamma\Lambda^{-1}\dot{\tilde{L}}\right) + 2\frac{\sigma}{\lambda}tr\left(\tilde{K}_f^T\Gamma\dot{\tilde{K}}_f\right) + 2\frac{\sigma}{\lambda}tr\left(\tilde{L}_f^T\Gamma\dot{\tilde{L}}_f\right) \tag{18}$$

From the error dynamics shown in Eq. (14), we can simplify it as:

$$\dot{V} = e^T\left(A_m^T\tilde{P} + \tilde{P}A_m\right)e + 2e^T\tilde{P}B_mL^{*-1}\left(-\tilde{K}y + \tilde{L}r\right) + 2tr\left(\tilde{K}^T\Gamma\Lambda^{-1}\dot{\tilde{K}}\right) \\ + 2tr\left(\tilde{L}^T\Gamma\Lambda^{-1}\dot{\tilde{L}}\right) + 2\frac{\sigma}{\lambda}tr\left(\tilde{K}_f^T\Gamma\dot{\tilde{K}}_f\right) + 2\frac{\sigma}{\lambda}tr\left(\tilde{L}_f^T\Gamma\dot{\tilde{L}}_f\right) + 2e^T\tilde{P}d(t) \tag{19}$$

Using Eqs. (7), (14) and (19) results as:

$$\dot{V} = -e^TQe - 2tr\left[\tilde{K}^T\Gamma\left(\left(B_m^T\tilde{P}ey^T - \sigma(\tilde{K} - \tilde{K}_f)\right) - \Lambda^{-1}\dot{\tilde{K}}\right)\right] \\ + 2tr\left[\tilde{L}^T\Gamma\left(\left(B_m^T\tilde{P}er^T + \sigma(\tilde{L} - \tilde{L}_f)\right) + \Lambda^{-1}\dot{\tilde{L}}\right)\right] + 2e^T\tilde{P}d(t) \tag{20}$$

for some arbitrary $Q = Q^T > 0$ in the Lyapunov equation.
Therefore, for the adaptive laws shown in Eq. (13), we have

$$\dot{V}\left(e, \tilde{K}, \tilde{L}, \tilde{K}_f, \tilde{L}_f\right) = -e^TQe + 2e^T\tilde{P}d(t) \leq -\lambda_{min}(Q)\|e\|^2 + 2\|e\|\lambda_{max}(\tilde{P})d_{max} \\ \leq -\lambda_{min}(Q)\|e\|\left(\|e\| - 2\frac{\lambda_{max}(\tilde{P})}{\lambda_{min}(Q)}d_{max}\right) \tag{21}$$

Consequently, the $\dot{V} < 0$ outside of the compact set given below:

$$E_0 = \left\{\left(e, \tilde{K}, \tilde{L}, \tilde{K}_f, \tilde{L}_f\right) : \|e\| \leq 2\frac{\lambda_{max}(\tilde{P})}{\lambda_{min}(Q)}d_{max} = e_0\right\} \tag{22}$$

The system trajectory $\left(e, \tilde{K}, \tilde{L}, \tilde{K}_f, \tilde{L}_f\right)$. starts outside the compact set, E_0 will reach the boundary of a compact set in a finite time and will remain there. Hence, $e, \tilde{K}, \tilde{L}, \tilde{K}_f, \tilde{L}_f \in L_\infty$

To find the L_2-norm of tracking error, we start with Eq. (21)

$$\dot{V}\left(e, \tilde{K}, \tilde{L}, \tilde{K}_f, \tilde{L}_f\right) = e^TQe + 2e^T\tilde{P}d(t) \leq -\lambda_{min}(Q)\|e\|^2 + 2\lambda_{max}(\tilde{P})\|e\|\|d\| \tag{23}$$

Using inequality regarding the completion of squares $2\|e\|\|d\| \leq \|e\|^2 + \|d\|^2$, we write Eq. (23) as:

$$\dot{V}\left(e, \tilde{K}, \tilde{L}, \tilde{K}_f, \tilde{L}_f\right) \leq \left(-\lambda_{min}(Q) + \lambda_{max}(\tilde{P})\right)\|e\|^2 + \lambda_{max}(\tilde{P})\|d\|^2 \tag{24}$$

Now integrating with respect to t and taking the limit as $t \to \infty$ results

$$\lim_{t\to\infty}\int_0^t \dot{V}d\tau \leq \left|-\lambda_{min}(Q) + \lambda_{max}(\tilde{P})\right|\lim_{t\to\infty}\int_0^t \|e\|^2 d\tau + \lambda_{max}(\tilde{P})\lim_{t\to\infty}\int_0^t \|d\|^2 d\tau \tag{25}$$

As we know

$$-\int_0^t \dot{V}d\tau = V\left(e_0, \tilde{K}_0, \tilde{L}_0, \tilde{K}_{f0}, \tilde{L}_{f0}\right) - V\left(e(t), \tilde{K}(t), \tilde{L}(t), \tilde{K}_f(t), \tilde{L}_f(t)\right) \\ \leq V\left(e_0, \tilde{K}_0, \tilde{L}_0, \tilde{K}_{f0}, \tilde{L}_{f0}\right) \tag{26}$$

With definition $V_0 \equiv V\left(e_0, \tilde{K}_0, \tilde{L}_0, \tilde{K}_{f0}, \tilde{L}_{f0}\right)$, the Eq. (25) becomes

$$\left|\lambda_{min}(Q) - \lambda_{max}(\tilde{P})\right| \|e\|_{L_2}^2 \leq V_0 + \lambda_{max}(\tilde{P}) d_{max}^2 \tag{27}$$

Then, L_2-norm of the tracking error is given by

$$\|e\|_{L_2}^2 \leq \frac{1}{\left|\lambda_{min}(Q) - \lambda_{max}(\tilde{P})\right|} \left(V_0 + \lambda_{max}(\tilde{P}) d_{max}^2\right). \tag{28}$$

The proof is completed.

Eq. (26) implies that $e(t)$ is (d_{max}^2)—small in the mean square sense (m.s.s.), that is, $e(t) \in S(d_{max}^2)$. Hence, tracking error is in the order of the disturbance only. When the external disturbance vanishes, the asymptotic property of the tracking error is guaranteed with the σ-modified adaptive laws using low-frequency learning. The performance of the MR-MRAC scheme will be identical to the MRAC if the external disturbances are zero. The inclusion of low-frequency learning in the adaptive laws of MRAC converts a pure integral type MRAC to a proportional-integral type MRAC [2]. The MR-MRAC enables fast learning ability and improves robustness. The time constant (i.e., λ^{-1}) of the low-pass filter needs to be sufficiently large enough to cut off the high-frequency oscillation [3]. In other words, λ needs to be sufficiently small. However, with very small values of λ, the gains $K(t), L(t)$ take a longer time to reach their steady-state values.

Figure 2 presents the schematics of the resulting MR-MRAC architecture Eq. (28). For the given reference input signal $r(t)$, the tracking error $e(t) \triangleq y(t) - y_m(t)$ is calculated using the plant output $y(t)$ and the reference model output $y_m(t)$. This tracking error is used in the σ-modified adaptive laws. The low-pass filter matrices $K_f(t)$ and $L_f(t)$ in the σ-modified adaptive laws are estimated from the low-frequency learning loop. The estimates of gain matrices $K(t)$ and $L(t)$ from the σ-modified adaptive laws are used in the controller to calculate control input array $u(t)$. This input array $u(t)$ is supplied to the APSP.

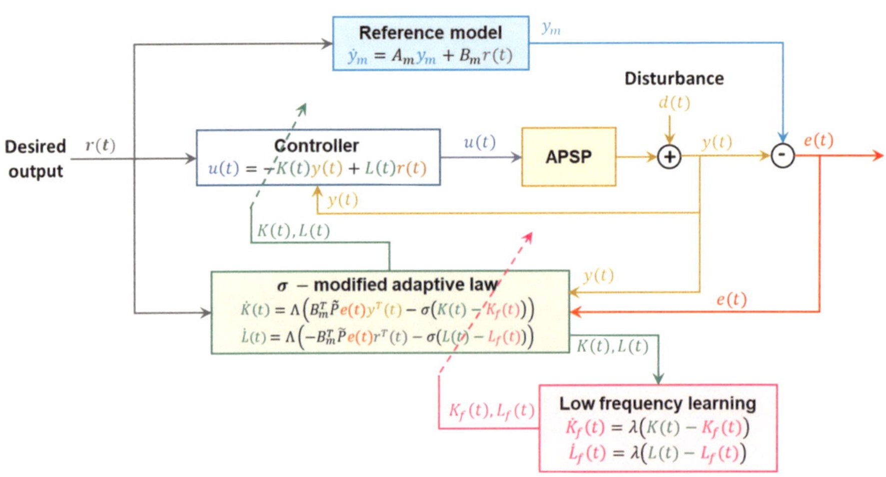

Figure 2.
Schematic of the MR-MRAC architecture [14].

4. Case studies

Shang et al. [8], among others, have provided a mathematical model for the APSP and the associated numerical model. We use LAVA-P software developed at the Idaho National Engineering and Environmental Laboratory to analyze the three-dimensional motions of powder particles within the plasma. **Figure 3** depicts the schematic of the proposed robust adaptive process control using the MR-MRAC for generating consistent quality ceramic coating using a single torch—single injection APSP system. Limits on the input variables with symbols indicated in parentheses are Ar flow rate (P), 20 slm $\leq P \leq$ 60 slm (standard liters per minute); H_2 flow rate (Q), 0 $\leq Q \leq$ 20 slm; and current (I), 300 A $\leq I \leq$ 600 A. The effect of disturbances is desired to die out within 50 ms of their occurrence. The sampling time of these simulations is 0.01 ms. The mean values of particles' axial velocity and temperature are computed in the 1-cm-wide window, 9.5 $\leq y \leq$ 10.5 cm located along the jet axis from the nozzle exit.

Linearizing the nonlinear dynamics of the mean axial velocity $v(t)$ and the mean temperature $T(t)$ around an equilibrium point results in the following multi-inputs and multi-outputs (MIMO) SS model:

$$\begin{Bmatrix} \dot{v}(t) \\ \dot{T}(t) \end{Bmatrix} = \begin{bmatrix} a_v & 0 \\ 0 & a_T \end{bmatrix} \begin{Bmatrix} v(t) \\ T(t) \end{Bmatrix} + \begin{bmatrix} b_{11} & b_{12} & b_{13} \\ b_{21} & b_{22} & b_{23} \end{bmatrix} \begin{Bmatrix} P(t) \\ Q(t) \\ I(t) \end{Bmatrix} + \begin{Bmatrix} d_v(t) \\ d_T(t) \end{Bmatrix}, \quad \begin{Bmatrix} v(0)=v_0 \\ T(0)=T_0 \end{Bmatrix} \tag{29}$$

where $d_v(t)$ and $d_T(t)$ are unknown smooth disturbances satisfying $\|d_v(t)\|_2 \leq d_{v,max}$, $\|d_T(t)\|_2 \leq d_{T,max}$, $\left\|\dot{d}_v(t)\right\|_2 \leq \dot{d}_{v,max}$, $\left\|\dot{d}_T(t)\right\|_2 \leq \dot{d}_{T,max}$ with positive bounds $d_{v,max}$, $d_{T,max}$, $\dot{d}_{v,max}$ and $\dot{d}_{T,max}$. Here, the parameters $a_v, a_T, b_{11}, \ldots, b_{23}$ are assumed as constants and are dependent on the equilibrium point. The reference model for the MR-MRAC is considered as follows:

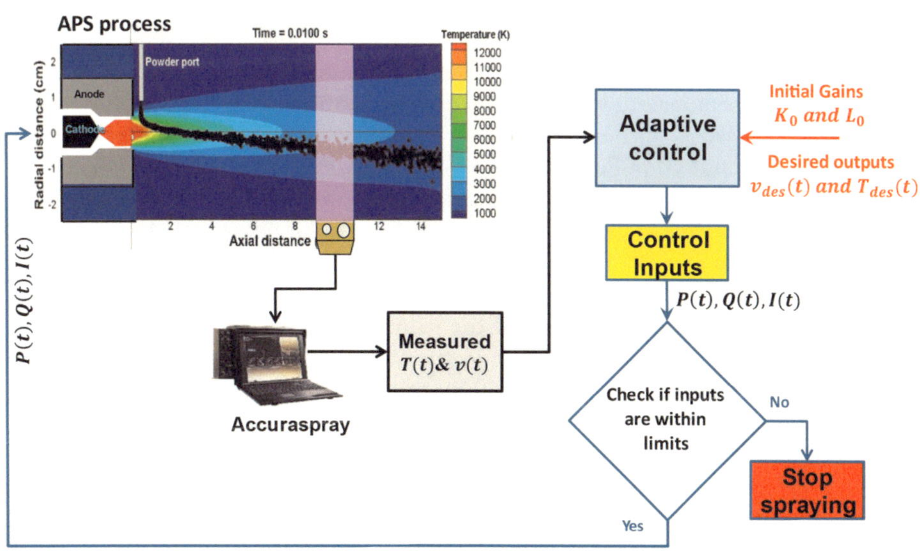

Figure 3.
Schematic of the proposed robust adaptive control scheme for an APSP [13].

$$\begin{Bmatrix} \dot{v_m}(t) \\ \dot{T_m}(t) \end{Bmatrix} = \begin{bmatrix} -0.5 & 0 \\ 0 & -0.5 \end{bmatrix} \begin{Bmatrix} v_m(t) \\ T_m(t) \end{Bmatrix} + \begin{bmatrix} 0.5 & 0 & 0.5 \\ 0 & 0.5 & 0.5 \end{bmatrix} \begin{Bmatrix} v_{des}(t) \\ T_{des}(t) \\ 0 \end{Bmatrix}, \quad \begin{Bmatrix} v_m(0) = v_0 \\ T_m(0) = T_0 \end{Bmatrix} \tag{30}$$

We choose $\tilde{P} = I_{2\times 2}$ and $l = 1$ in Eq. (13).

The objective is to force the measured MPSs $v(t)$ and $T(t)$ from the LAVA-P to track the reference outputs $v_m(t)$ and $T_m(t)$ and to convergence to $v_{des}(t)$ and $T_{des}(t)$, respectively, within a small bound and within the settling time = 50 ms despite changes in the noise variables. The controller adjusts inputs, such as the argon flow rate $P(t)$, the hydrogen flow rate $Q(t)$, and the current $I(t)$ using control law in Eq. (5), σ-modified adaptive laws in Eq. (13), and low-pass filter in Eq. (10). From Ref. [18], we choose the following initial gains in Eqs. (10) and (13):

$$K_0 = \begin{bmatrix} 0 & 0 \\ 0 & 0 \\ 0 & 0 \end{bmatrix}; \quad L_0 = \begin{bmatrix} 0.0004 & 0.0123 & 0 \\ 0.0001 & 0.0031 & 0 \\ 0.0045 & 0.1430 & 0 \end{bmatrix}. \tag{31}$$

4.1 Performance comparison of MRAC, R-MRAC, and MR-MRAC

For this study, we use the following adaptive gain matrix in Eq. (13):

$$\Lambda = \gamma \times \begin{bmatrix} 10^{-8} & 0 & 0 \\ 0 & 10^{-10} & 0 \\ 0 & 0 & 10^{-9} \end{bmatrix} \tag{32}$$

where γ is the adaptive constant introduced as tuning parameters to achieve the desired transient responses. First, the effect of the adaptive constant γ of the standard MRAC (i.e., MR-MRAC with $\sigma = 0$ and $\lambda = 0$) is investigated using arbitrarily chosen values of γ as 1, 5, and 20 in Eq. (32) under the step variation of average injection velocity. The variations of this disturbance and the corresponding control responses are presented in the first column of **Figure 4**. While the larger values of γ speed up the adaption rate for each input; however, it resulted in high-frequency oscillations in control responses, and the performance of the MRAC became unstable. The smaller value of γ limits the convergence rate of the plant responses; thus, it takes a longer time to reach the desired values. For $\gamma = 5$, the mean particles' states have reached the desired values within a settling time of 50 ms with a few minor oscillations in the control responses.

The second column of **Figure 4** depicts the effect of damping parameter σ in the R-MRAC (i.e., MR-MRAC with $\lambda = 0$) for σ = 10^4, 10^6, and 10^8. To investigate the fast adaption capability of the proposed controller, the high adaptive constant $\gamma = 20$ is considered for this exercise. If the damping parameter σ increases from 10^4 to 10^6, the frequency of oscillations in the control responses is reduced; however, the steady-state error between the desired and the measured mean particles' states is increased. For

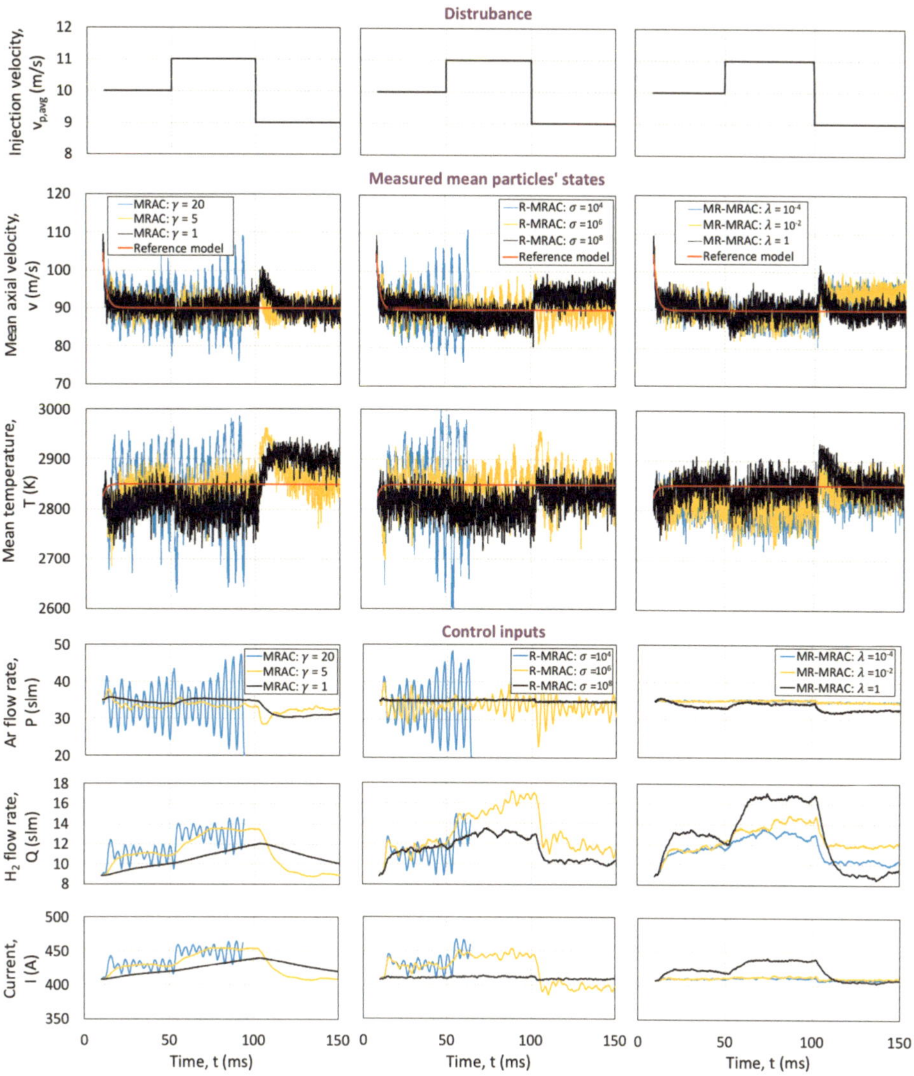

Figure 4.
Effect of adaptive constant γ in MRAC (first column), damping constant σ in RMRAC (second column for $\gamma = 20$), and filter constant λ in MR-MRAC (third column for $\gamma = 20$ and $\sigma = 10^8$).

$\sigma = 10^8$, the effect of damping is very high, and this leads to an undesirable steady-state error between the measured and desired outputs.

The performance of the MR-MRAC is investigated for $\gamma = 20$ and $\sigma = 10^8$ with the filter constants $\lambda = 10^{-4}$, $\lambda = 10^{-2}$, and $\lambda = 1$. The fast adaption using a high-gain learning rate is achieved with the MR-MRAC scheme for $\lambda = 1$. From the third column of **Figure 4**, we can observe that the high-frequency oscillations with $\gamma = 20$ and steady-state error for $\sigma = 10^8$ have been reduced with MR-MRAC and the effect of disturbance is attenuated within 20 milliseconds. The smooth variations in the mean particles' states and control inputs are acquired. This illustrates the point that the designed MR-MRAC achieves the fast and robust adaption without creating highfrequency oscillations or steady-state errors.

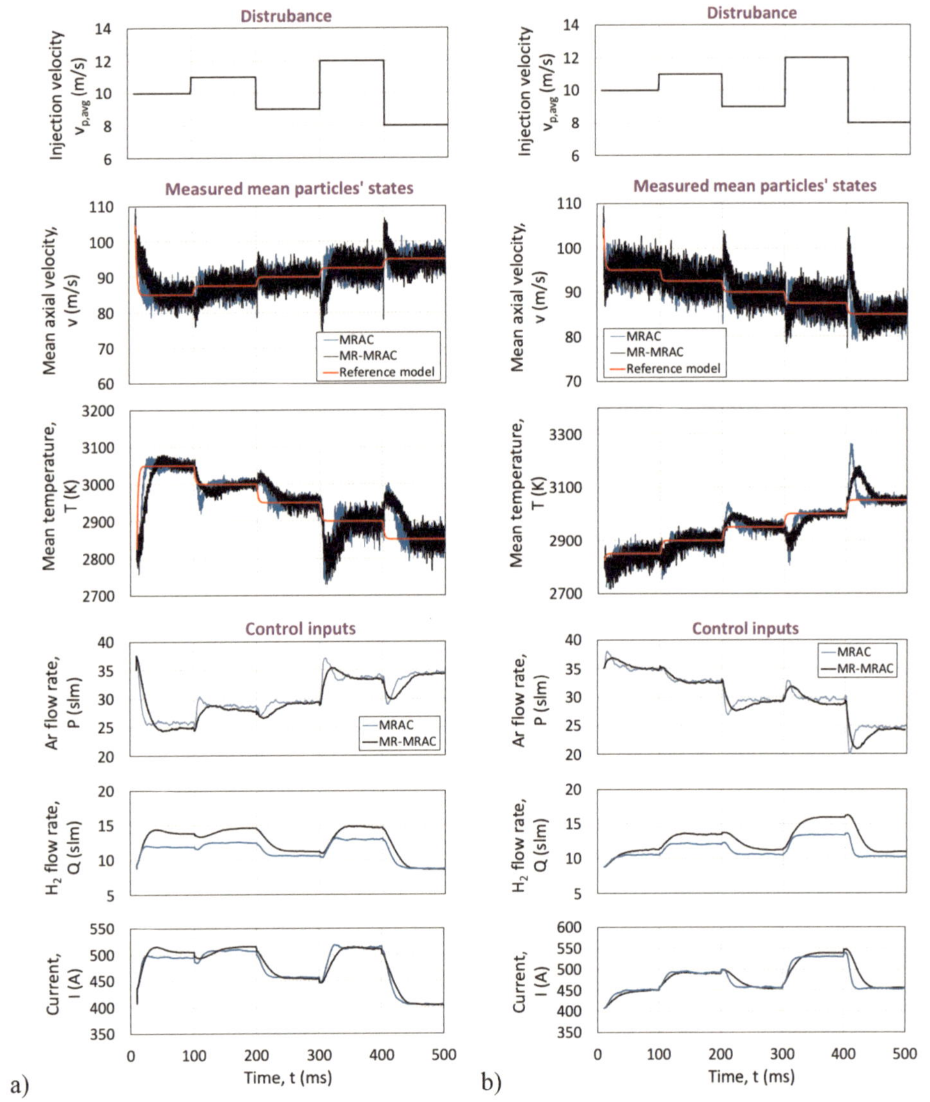

Figure 5.
The closed-loop control responses of MRAC and MR-MRAC for a) increasing desired mean particles' axial velocity and decreasing desired mean particles' temperature; and b) the decreasing desired mean particles' axial velocity but increasing desired mean particles' temperature in the presence of step disturbance in average injection velocity.

Therefore, the performance of the MR-MRAC is superior compared to that of the standard MRAC.

4.2 Effect of disturbance variations on performance of MR-MRAC

First, we now check the performance of the MR-MRAC for adaptive tracking when the desired values of mean particles' states vary with time in the presence of a variation of the disturbance. The following adaptive gain matrix is used for the rest of the analyses:

$$\Lambda = \begin{bmatrix} 2\times 10^{-8} & 0 & 0 \\ 0 & 2\times 10^{-10} & 0 \\ 0 & 0 & 5\times 10^{-9} \end{bmatrix} \tag{33}$$

Here, the damping constant $\sigma = 10^8$ and the filter constant $\lambda = 1$. In this case, the performances of the MRAC and the MR-MRAC are compared, and the subsequent results are presented in **Figure 5** for step variations in $v_{des}(t)$ *and* $T_{des}(t)$ in the presence of the step variation of average particles' injection velocity. It is evident that both controllers effectively force the mean particles' states to follow the outputs of the reference model despite larger magnitudes of the error between the measured and the desired particles' states. However, the magnitude of the mean temperature oscillations is significantly dampened using the MR-MRAC. Even if the mean particles' states reach the desired values faster than the MR-MRAC in this particle case, faster convergence can be achieved for MR-MRAC by further fine-tuning of the filter constant λ.

The effect of the simultaneously varied disturbances, such as the average injection velocity of particles and the arc voltages (in the first and the second row of **Figure 6**), on the mean axial velocity and mean temperature are shown in the third and the fourth rows of **Figure 6**, respectively. The performance of the MR-MRAC is investigated under these disturbances, and the varied control outputs are shown in the fifth and the sixth rows of **Figure 6**, and the corresponding control inputs are presented in the remaining rows of **Figure 6**. These examples establish the effectiveness of the designed process controller in mitigating the effects of various classes of disturbances for generating ceramic coatings using the APSP. Of course, in practice, disturbances are not limited to those stipulated here.

5. Conclusion

This chapter proposes two modifications to the standard model reference adaptive control (MRAC) to improve the robustness and the fast adaption, and its application to the atmospheric plasma spray process (APSP) to improve repeatability and reproducibility of coating quality and simultaneously decreasing manufacturing costs. The MRAC lacks robustness in the presence of process uncertainties and bounded external disturbances, which is a well-known drawback. This issue is handled by the modified robust MRAC (MR-MRAC) scheme, which consists of sigma-modified adaptive laws of MRAC with a low-pass weight estimated filter. The control objective for the APSP is to get consistency in mean particles' states, such as mean axial velocity and mean temperature, before impacting the substrate. The performance of this strategy is tested on numerical software LAVA-P simulates the APSP. The MR-MRAC shows better performance compared to standard MRAC under external disturbances. The desired consistency in mean particles' states is achieved despite artificially induced disturbances by varying control inputs through fast adaption. The smooth variations in the mean particles' states and control inputs are acquired without generating high-frequency oscillations or steady-state errors.

We anticipate that the MR-MRAC will perform equally well in practical applications and economically enable the production of high-quality coatings. This control architecture also is feasible for other coating methods, such as the HVOF spray process, physical vapor deposition, and chemical vapor deposition.

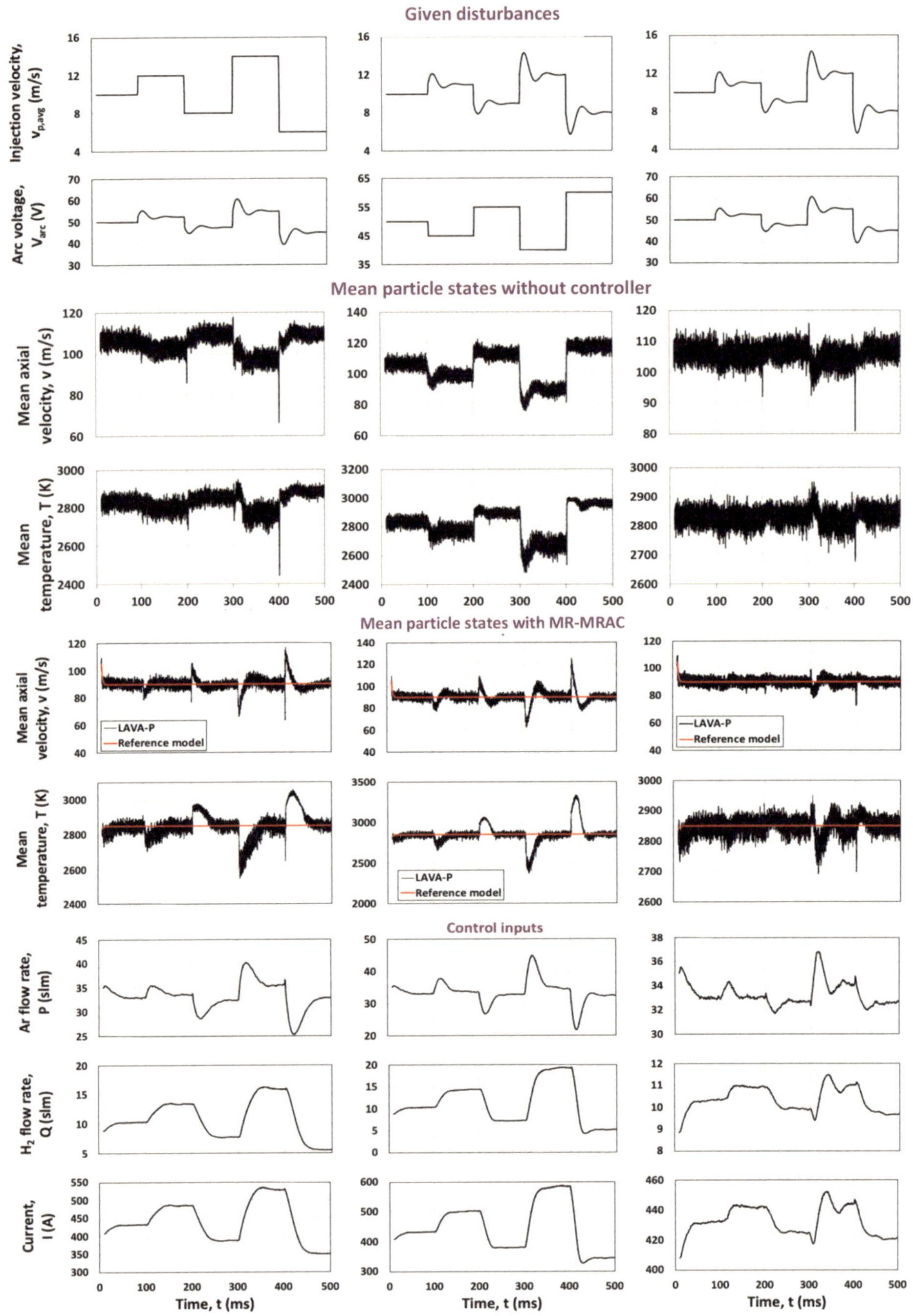

Figure 6.
The closed-loop responses of the MR-MRAC system in the presence of combined disturbances of the average powder injection velocity and the arc voltage.

Conflict of interest

The authors declare that they have no conflict of interest.

Author details

Balachandar Guduri
Virginia Tech Transportation Institute, Virginia Polytechnic Institute and State University, Blacksburg, VA, USA

*Address all correspondence to: gbalu@vt.edu

References

[1] Tao G. Multivariable adaptive control: A survey. Automatica. 2014;**50**(11): 2737-2764

[2] Yucelen T, Haddad WM. Low-frequency learning and fast adaptation in model reference adaptive control. IEEE Transactions on Automatic Control. 2012;**58**(4):1080-1085

[3] Egardt B. Stability of Adaptive Controllers. Berlin Heidelberg: Springer; 1979

[4] Rohrs C et al. Robustness of continuous-time adaptive control algorithms in the presence of unmodeled dynamics. IEEE Transactions on Automatic Control. 1985;**30**(9):881-889

[5] Ioannou PA, Sun J. Robust adaptive control. 1: Upper Saddle River, NJ: PTR Prentice-Hall; 1996

[6] Shi Z, Zhao L. Robust model reference adaptive control based on linear matrix inequality. Aerospace Science and Technology. 2017;**66**:152-159

[7] Sampath S et al. Thermal spray processing of FGMs. MRS Bulletin. 1995; **20**(1):27-31

[8] Shang S et al. Effect of turbulence modulation on three-dimensional trajectories of powder particles in a plasma spray process. Journal of Physics D: Applied Physics. 2014;**47**(40):405206

[9] Williamson R, Fincke J, Chang C. A computational examination of the sources of statistical variance in particle parameters during thermal plasma spraying. Plasma Chemistry and Plasma Processing. 2000;**20**(3):299-324

[10] Wan Y, et al., Model and Powder Particle Heating, Melting, Resolidification, and Evaporation in Plasma Spraying Processes. 1999. p. 691-699

[11] Fincke JR et al. Diagnostics and control in the thermal spray process. Surface and Coatings Technology. 2001; **146**:537-543

[12] Westergård R et al. The erosion and abrasion characteristics of alumina coatings plasma sprayed under different spraying conditions. Tribology International. 1998;**31**(5):271-279

[13] Guduri B et al. Adaptive process control for achieving consistent particles' states in atmospheric plasma spray process. SN Applied Sciences. 2021;**3**(3): 1-21

[14] Guduri B, Batra R. Robust model reference adaptive controller for atmospheric plasma spray process. SN Applied Sciences. 2022;**4**(4):1-10

[15] Guduri B, Batra RC. Adaptive control of the atmospheric plasma spray process for functionally graded thermal barrier coatings. Advances in Materials Science and Engineering. 2022; **2022**

[16] Ioannou PA, Datta A. Robust adaptive control: A unified approach. Proceedings of the IEEE. 1991;**79**(12): 1736-1768

[17] Ioannou PA, Kokotovic PV. Instability analysis and improvement of robustness of adaptive control. Automatica. 1984;**20**(5):583-594

[18] Guduri B. Adaptive Process Control for Achieving Consistent Mean Particles' States in Atmospheric Plasma Spray Process. US: Virginia Tech; 2022

Chapter 2

Position Control of External Rotor Permanent Magnet Synchronous Motor with Model Predictive Control

Goksel Kizir and Abdul Furkan Kanburoglu

Abstract

Permanent magnet synchronous motors (PMSM) are widely used in applications such as industrial, automotive, robotics, medical, and military applications. External rotor permanent magnet synchronous motors (ER-PMSM), a type of PMSM, provide high output torque at low volume. Although potentially usable in robotic and military applications, they are often used in drones due to the cogging torques involved in the output torques. In this chapter, using Model Predictive Control (MPC) and PID controller, position control of ER-PMSMs will be presented. The controlled autoregressive and integrated moving average (CARIMA) model is determined for the MPC algorithm. Experimental studies are carried out with these position control structures. In addition, the selected controllers will be compared using different position references. Their advantages and disadvantages will be explained in detail.

Keywords: model predictive control, external rotor permanent magnet synchronous motor, field oriented control, controlled autoregressive and integrated moving average, position control

1. Introduction

Model predictive control is frequently preferred in industrial, automotive, and military applications recently as a control method. The main reason for this is the desire to control the system in the most optimal way. If there is a model to represent the system, optimum control signals can be produced with model predictive control.

One of the application areas of MPC is the control of electrical machines. The definition of electrical machines is a device that converts electrical energy into mechanical energy. Permanent Magnet Synchronous Motors (PMSMs), which make this conversion in small volumes and high efficiency, are one of the most preferred types of electrical machines. PMSMs are used in many industrial, automotive, robotic, and military applications. There are many methods to control PMSMs with and without sensors in the literature [1–7].

Two areas where PMSMs are frequently preferred for similar control purpose are robotics and military applications. PMSMs are used as actuators that control a joint in robotic applications, as well as control the angular position of a vane in military applications. This position control may need to be of high precision in both areas. Since position data is needed as feedback in position control of PMSMs, sensorless control methods cannot be applied in these applications. In such applications, the control methods preferred recently are intelligent control methods or model-based control methods [8–12].

Thanks to the developing technology and developing production infrastructures, PMSMs in different structures can be produced. Coreless PMSMs, a type of PMSM, can produce constant torque. The stator windings of Coreless PMSMs are air wound instead of a core. In this way, much lower cogging torque occurs at output torques compared to other PMSMs. Due to their low cogging torque, they are preferred in robotics and military applications [13–16].

The main drawback of these coreless PMSMs is their low output torque. In order to meet high output torque requirements for relatively small actuator designs, higher transmission ratios may be considered. However, designing an actuator with higher transmission ratios arise other considerations such as efficiency, backlash, etc.

ER-PMSM has gradually become popular with the characteristics of low loss, simplistic-compact structure, and reliable in operation. Unlike Coreless PMSMs, ER-PMSMs can provide high output torques at a similar volume. Since the stator windings of most of the ER-PMSMs have cores, cogging occurs in the output torques of the ER-PMSMs. This cogging torque causes a disruption effect on the control. There are studies in the literature for modeling/reducing the cogging torque [17–20]. However, ER-PMSMs do not have position control application. In addition, most of the developed ER-PMSMs are used in drones.

In this study, which aims to produce high output torque with low transmission ratio, the use of ER-PMSMs as an actuator in an application that requires precise position control is evaluated. It is planned to increase the efficiency of the transmission mechanism with a low transmission ratio. In addition, with the position controller to be designed, it is desired to minimize the effect of the cogging torque on the control.

MPC is selected for position control. Additionally, a comparison is added in order to evaluate the performance of MPC. The sequential PID control structure, which is frequently preferred for motor control in the literature, is preferred for this comparison. In the third part, there are four different test scenario results of two controllers. The fourth section is the section where the test results are discussed. In the last part, the results of the study are evaluated.

2. PMSM model and control strategies

The block diagram of PMSM and control strategies in this chapter is given in **Figure 1**. A three-phase PMSM is driven by a PWM inverter circuit. Three-phase control voltage is applied to the PWM inverter circuit, and the three-phase current value is read by analog sensors in this circuit. In addition, the angular position of the PMSM is read by an incremental encoder.

In the control algorithm, the rotor position of PMSM is controlled by both MPC and PID. Both control structures calculate the PMSM q-axis control voltage. While MPC only needs angular position data for this calculation, PID controller structure

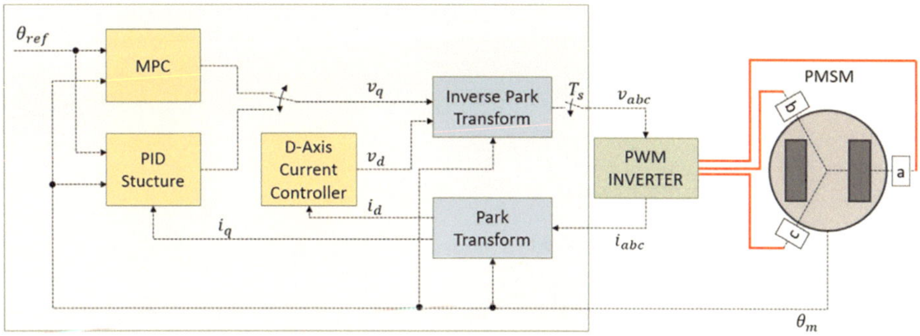

Figure 1.
PMSM and control strategies block diagram.

also needs the q-axis current of the PMSM. In addition to the position control, the d-axis current of the PMSM is kept at zero by a different controller. In this chapter, PID controller structure is chosen as the d-axis current controller.

Obtained d- and q-axis voltages are converted to PMSM phase voltages by Inverse Park Transform. In addition, the d- and q-axis currents required by the PID controller, and the d-axis current controller are obtained by Park Transform.

In the second part, PMSM model and control strategies are explained. Field oriented control (FOC), which is the PMSM driving method, is explained, the PID controller structure and the selection criteria of the PID controller coefficients are mentioned. Finally, the determined MPC method is explained in detail.

2.1 PMSM model

PMSMs are synchronous motors with permanent magnets on their rotors and windings on their stators [7]. The stator windings are usually three-phases, although five- or eight-phases configurations are also available. Considering the positions of their rotors, PMSMs can be classified in two ways as external (outer) rotor and inner rotor.

Although this difference in design changes the speed and torque characteristics of both motors, their mathematical models are the same. PMSM mathematical models are available for both three phases (abc-phases) and two axes (dq-axis). Since PMSM control can be handled linearly with two-axis modeling, two-axis modeling is preferred in this study. The equations of the PMSM according to the dq-axis set are as follows.

$$v_d = R_s i_d - \omega L_q i_q + \rho\left(L_d i_d + \lambda_{pm}\right) \tag{1}$$

$$v_q = R_s i_q + \omega\left(L_d i_d + \lambda_{pm}\right) + \rho L_q i_q \tag{2}$$

$$T_e = 3p/2\left(\lambda_{pm} i_q + \left(L_d - L_q\right) i_d i_q\right) \tag{3}$$

$$T_e - T_l - T_{cogg}(\theta_m) = \rho\omega_m J + B\omega_m \tag{4}$$

R_d and R_q parameters in the equations represent the relevant axis resistances. L_d and L_q represent the relevant axis inductances. ω is electrical where ω_m is mechanical velocity. p is the number of pole pairs. B is the friction coefficient, λ_{pm} is magnet mutual flux linkage. T_e is electrical torque generated by the motor and T_{cogg} is the

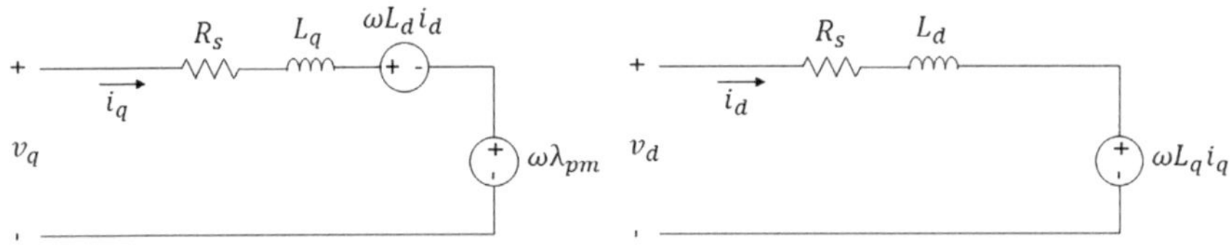

Figure 2.
Equivalent circuit of PMSM electrical equations.

cogging torque depending on the motor position θ_m. J is inertia of rotor. ρ is the derivative operator. The equivalent circuit of PMSM electrical equations is as in **Figure 2**.

While controlling PMSMs, control voltage is applied to q-axis, and a voltage is applied to the d-axis in order to keep the d-axis current at zero. Assuming the d-axis current is controlled as zero, Eqs. (2) and (3) simplify as in Eqs. (5) and (6).

$$v_q = R_s i_q + \omega\lambda_{pm} + \rho L_q i_q \tag{5}$$

$$T_e = 3p/2(\lambda_{pm} i_q) \tag{6}$$

PMSM transfer function to be created using Eqs. (4)–(6) will be as in **Figure 3**.

In MPC, the PMSM mathematical equations that are arranged for the motor position are the output and the q-axis voltage is the input of the system. The transfer function in MPC is given in Eq. (7).

$$G_{pmsm} = \frac{3/2 * p * \lambda_{pm}}{L_q J s^3 + (R_s J + L_q B)s^2 + \left(3/2(p\lambda_{pm})^2 + R_s B\right)s} \tag{7}$$

While obtaining this equation, it is assumed that current flow through the d-axis is zero. Hence the PMSM mathematical equations are simplified. The d-axis current of the PMSM is kept at zero with a PID controller.

2.2 Field oriented control

The Field Oriented Control (FOC) is a method based on controlling the stator currents represented by vectors. FOC gives proficient control capability over the full torque and speed ranges. Its implementation requires transformation of stator currents from the stationary reference frame to the rotor flux reference frame (also

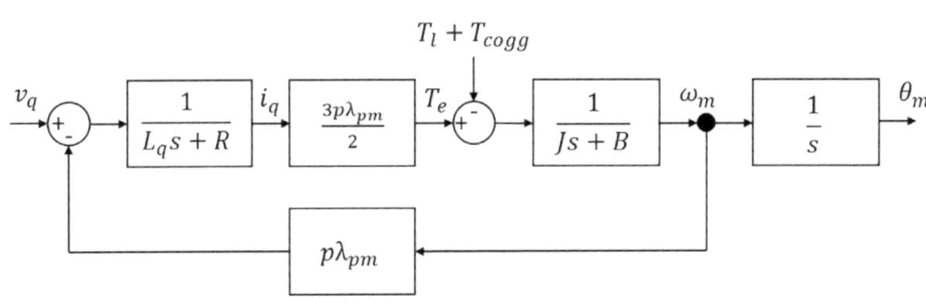

Figure 3.
PMSM transfer function block diagram.

known as *d-q* reference frame). It was originally developed for high-performance motor applications that are required to operate smoothly.

Field oriented controlled machines need two components as input references: the torque component (aligned with the q-axis) and the flux component (aligned with d-axis). The torque reference of the PMSM is controlled by q-axis. The PMSM flux is kept constant by controlling of the d-axis. Besides, d-axis is used for parameter estimation in some applications [9].

The Park Transformation is a transformation between phases of the PMSM and axis of the FOC. The Park transformation converts abc-phases of the PMSM to dq-axis of the FOC and The Inverse Park Transformation coverts dq-axis of the FOC to abc-phases of the PMSM. Park's Transformation $s_{dq} = T^{dq}_{abc}(\theta)s_{abc}$ and inverse Park's transformation $s_{abc} = T^{abc}_{dq}(\theta)s_{dq}$ is given in Eqs. (8) and (9) respectively where θ is electrical angle.

$$T^{dq}_{abc}(\theta) = \frac{2}{3}\left[\cos(\theta)\ \cos(\theta-2\pi/3)\ \cos(\theta+2\pi/3)\ \sin(\theta)\ \sin(\theta-2\pi/3)\ \sin(\theta+2\pi/3)\right] \quad (8)$$

$$T^{qbc}_{dq}(\theta) = \frac{2}{3}\left[\cos(\theta)\ \sin(\theta)\ \cos(\theta-2\pi/3)\ \sin(\theta-2\pi/3)\ \cos(\theta+2\pi/3)\ \sin(\theta+2\pi/3)\right] \quad (9)$$

2.3 PID controller

The classical PID controller transfer function is as in Eq. (10). The PID controller structure is shown in **Figure 4**.

$$G_{pid} = K_p + \frac{K_i}{s} + K_d s \quad (10)$$

The PID control structure used in this study is shown in **Figure 4**. A cascade structure is preferred as the PID control structure. The output of the position controller is set as the velocity controller reference where the output of the velocity controller is the q-axis current controller reference. The d-axis current reference is set to zero. PMSM position and currents are measured from the system. The velocity value is obtained by taking the discrete derivative of the measured position value.

PID controller outputs are produced on dq-axis and applied to three phases of PMSM with Inverse Park Transform. Utilizing the current values measured from the three phases of the PMSM, the currents in the dq-axis are calculated with Park Transform (**Figure 5**).

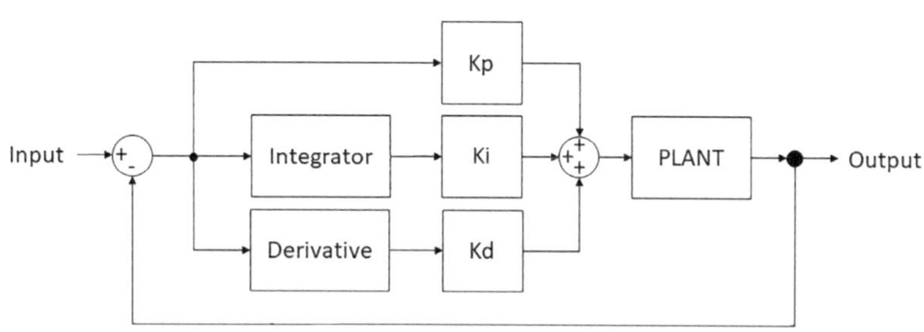

Figure 4.
Block diagram of PID Controller.

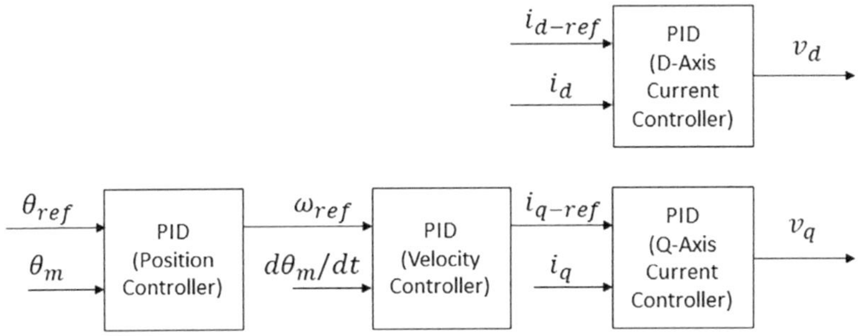

Figure 5.
Block diagram of PID controller structure.

The D coefficient of the d- and q-axis current controllers in the sequential PID controller structure has been determined as zero. This is because current measurements are made analog and the derivative expression of current noise degrades the controller performance. Eq. (11) is used while determining the PI controller coefficients of these two current controllers. Eq. (11) is simplified version of Eqs. (1) and (2) and has a siso structure. Since the electrical characteristic of PMSM is much faster than the mechanical characteristic, there is no velocity dependent expression in Eq. (11).

$$\frac{i_x}{v_x} = G_{cur\_pid} = \frac{1}{L_x s + R_s} \tag{11}$$

In the velocity controller, as in the current controllers, the D coefficient is set to zero. The reason for this is that the velocity is not measured directly in the system and is obtained by taking the derivative of the position data. The second derivative of the position data degrades the controller performance. Eq. (12) is used while determining the PI controller coefficients of the speed controller. In addition, the velocity controller bandwidth is determined to be 1/10 of the current controller bandwidth. Since the current controller is 10 times more than the speed controller, changes in the current dynamics according to the Nyquist Criteria will not affect the dynamics of the speed controller.

$$\frac{\omega_m}{i_q} = G_{vel\_pid} = \frac{3p/2(\lambda_{pm})}{Js + B} \tag{12}$$

In the position controller, the P controller is selected. In addition, the position controller bandwidth is determined to be 1/10 of the velocity controller bandwidth. Thus, the velocity of the PMSM will be 10 times faster than the dynamics of the position controller, and only the integral expression will remain between the speed and the position. Eq. (13) is used to determine the position controller P coefficient. The I term of PID is used to prevent steady state error in current and speed controllers. In addition, there is no steady state error inherent in the position controller designed equation. Therefore, choosing the position controller as P instead of PID will provide ease of operation.

$$\frac{\theta_m}{\omega_m} = G_{pos\_pid} = \frac{1}{s} \tag{13}$$

2.4 Model predictive control

The controlled autoregressive and integrated moving average (CARIMA) model is determined as t MPC algorithm [21]. The reason for choosing the CARIMA model as the MPC method is not only the optimization process can be made offline but also, in this method, the processing load on the microcontroller is relatively low. Generic CARIMA equation is given in Eq. (14).

$$a(z)y_k = b(z)u_k + d_k \tag{14}$$

$a(z)$ and $b(z)$ in Eq. (14) represent the transfer function of the system, and d_k represents disturbance and measurement noise effects simultaneously.

Unlike the generic CARIMA model, noise and disturbances are removed and the simplest transfer function model is obtained as in Eq. (15).

$$a(z)y_k = b(z)u_k \tag{15}$$

Eq. (16) is obtained if both sides of the Eq. (15) are multiplied by the expression Δ.

$$A(z)y_k = b(z)(\Delta u_k) \tag{16}$$

$$\mathrm{a(z)}\Delta = \mathrm{A(z)} \tag{17}$$

If the Eq. (16) is written in expanded and matrix form for n steps, Eq. (18) is obtained.

$$C_A \begin{bmatrix} y_{k+1} \\ y_{k+2} \\ \vdots \\ y_{k+n} \end{bmatrix} + H_A \begin{bmatrix} y_k \\ y_{k-1} \\ \vdots \\ y_{k-n+1} \end{bmatrix} = C_b \begin{bmatrix} \Delta u_k \\ \Delta u_{k+1} \\ \vdots \\ \Delta u_{k-m} \end{bmatrix} + H_b \begin{bmatrix} \Delta u_k \\ \Delta u_{k-1} \\ \vdots \\ \Delta u_{k-m+1} \end{bmatrix} \tag{18}$$

The coefficients in Eq. (18) are as follows.

$$C_A = \begin{bmatrix} 1 & 0 & 0 & \cdots & 0 \\ A_1 & 1 & 0 & \cdots & 0 \\ A_2 & A_1 & 1 & \cdots & 0 \\ \vdots & \vdots & \vdots & \ddots & \vdots \\ A_{n-1} & A_{n-2} & A_{n-3} & \cdots & 1 \end{bmatrix} \tag{19}$$

$$H_A = \begin{bmatrix} A_1 & A_2 & \cdots & A_{n-1} & A_n \\ A_2 & A_3 & \cdots & A_n & 0 \\ \vdots & \vdots & \ddots & \vdots & \vdots \\ A_{n-1} & A_n & \cdots & 0 & 0 \\ A_n & 0 & \cdots & 0 & 0 \end{bmatrix} \tag{20}$$

$$C_b = \begin{bmatrix} b_1 & 0 & 0 & \cdots & 0 \\ b_2 & b_1 & 0 & \cdots & 0 \\ b_3 & b_2 & b_1 & \cdots & 0 \\ \vdots & \vdots & \vdots & \ddots & \vdots \\ b_m & b_{m-1} & b_{m-2} & \cdots & b_1 \end{bmatrix} \tag{21}$$

$$H_b = \begin{bmatrix} b_2 & b\_3 & \cdots & b_{m-1} & b_n \\ b_3 & b_4 & \cdots & b_m & 0 \\ \vdots & \vdots & \ddots & \vdots & \vdots \\ b_{m-1} & b_m & \cdots & 0 & 0 \\ b_m & 0 & \cdots & 0 & 0 \end{bmatrix} \tag{22}$$

Eq. (23) is obtained if Eq. (18) is arranged to give the estimation of the output value. The expression $\vec{y}_{k+1}$ in Eq. (23) is the estimation of the output value in the next step. $\vec{u}_k$ value is the input signal to be applied to the system in current step. $\vec{u}_{k-1}$ is the input signal applied to the system in the previous step and $\vec{y}_k$ is the output value in the current step.

$$\vec{y}_{k+1} = H\Delta\vec{u}_k + P\Delta\vec{u}_{k-1} - Q\vec{y}_k \tag{23}$$

The coefficients in Eq. (23) are as follows.

$$H = C_A^{-1}C_b \tag{24}$$

$$P = C_A^{-1}H_b \tag{25}$$

$$Q = C_A^{-1}H_A \tag{26}$$

The transfer function to be used for MPC is obtained in Eq. (23). In the next step, the cost function of the system should be determined. The cost function used in this study is given in Eq. (27).

$$J = \Delta\vec{u}_k^T S\Delta\vec{u}_k + \Delta\vec{u}_k^T 2X \begin{bmatrix} \Delta\vec{u}_{k-1} \\ \vec{y}_k \\ \vec{y}_{k+1} \end{bmatrix} \tag{27}$$

The coefficients in Eq. (27) are as follows.

$$S = H^T W_y H + W_u \tag{28}$$

$$X = \left[H^T W_y P \; H^T W_y Q \quad -H^T W_y \right] \tag{29}$$

If the matrices in Eqs. (28) and (29) are updated to calculate the next step of the prediction, instead of n step long vector, the calculations are simplified as follows.

$$S = H_{1xn}^T W_y H_{1xn} + W_u \tag{30}$$

$$X = \left[H_{1xn}^T W_y P \quad H_{1xn}^T W_y Q \quad -H_{1xn}^T W_y \right] \tag{31}$$

According to the cost function, the solution is given in Eq. (32).

$$\Delta\vec{u}_k = P_r\vec{y}_{k+1} - D_k\Delta\vec{u}_{k-1} - N_k\vec{y}_k \tag{32}$$

$$P_r = S^{-1}H^T W_y \tag{33}$$

$$D_k = S^{-1}H_{1xn}^T W_y P \tag{34}$$

$$N_k = S^{-1}H_{1xn}^T W_y Q \tag{35}$$

The block diagram of MPC is as in **Figure 4**. To obtain the control signal, the reference input value is multiplied by the P_r matrix. In next step, the output measurement of the system and the measurement matrix obtained from the previous output values are multiplied by the D_k matrix. In final step, the delta of the reference signal applied to the system and the matrix obtained from the previous values of these deltas are multiplied by the N_k matrix. The result obtained is added together. This value is the delta reference value to be applied to the system. This delta reference value is added at each step (**Figure 6**).

In order to convert the PMSM transfer function in Eq. (7) to CARIMA form in Eq. (11), it is necessary to switch from the s transform of Eq. (36) to the z transform. Tustin method is selected for this process. According to the Tustin method, the following equation is used to obtain the z transform of an equation in the s transform. T_s is the sampling time of the z transform.

$$s \leftarrow \frac{2}{T_s}\frac{z-1}{z+1} \tag{36}$$

The CARIMA form of the PMSM transfer function is as in Eqs. (38) and (39).

$$s \leftarrow \frac{2}{T_s}\frac{z-1}{z+1} \tag{37}$$

$$a(z) = a_3z^3 + a_2z^2 + a_1z + a_0 \tag{38}$$

$$b(z) = \frac{3}{2}p\lambda_{pm}T_s^3\left(z^3 + 3z^2 + 3z + 1\right) \tag{39}$$

$$a_3 = \left(4T_S(R_sJ + L_qb) + 8L_qJ + 2T_s^2\left(\frac{3}{2}\left(p\lambda_{pm}\right)^2 + R_sb\right)\right) \tag{40}$$

$$a_2 = \left(-4T_S(R_sJ + L_qb) - 24L_qJ + 2T_s^2\left(\frac{3}{2}\left(p\lambda_{pm}\right)^2 + R_sb\right)\right) \tag{41}$$

$$a_1 = \left(-4T_S(R_sJ + L_qb) + 24L_qJ - 2T_s^2\left(\frac{3}{2}\left(p\lambda_{pm}\right)^2 + R_sb\right)\right) \tag{42}$$

$$a_0 = \left(4T_S(R_sJ + L_qb) - 8L_qJ - 2T_s^2\left(\frac{3}{2}\left(p\lambda_{pm}\right)^2 + R_sb\right)\right) \tag{43}$$

After obtaining the CARIMA form, the other matrices required for MPC are calculated in MATLAB because their sizes are very large.

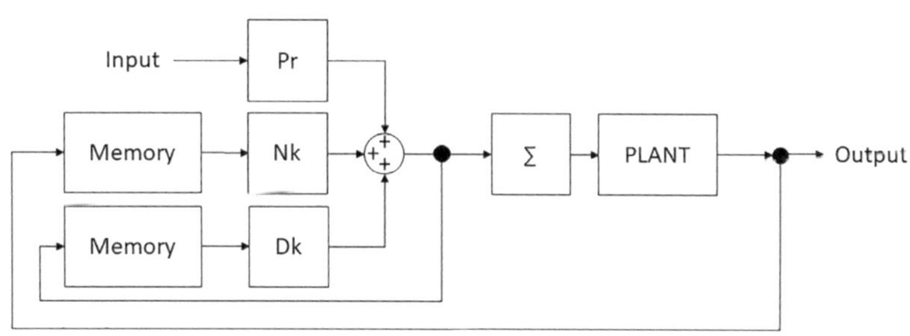

Figure 6.
Block diagram of MPC.

3. Experimental results

The test system is given in **Figure 7**. In the test system, there is an absolute encoder connected to the output shaft of the ER-PMSM for measuring the position of the ER-PMSM. The resolution of this encoder is 21 bits. This value is sufficient for us to observe the movement precisely. There is a driver card that controls and drives the ER-PMSM. FPGA is used as a smart unit in this card. These three components are used in all of the no-load tests of the ER-PMSM described below.

Experimental studies are carried out with a driver card that is controlled by FPGA. Since the calculations can be made in parallel and quickly, control performance can be evaluated without being affected by the computation load and time.

The load motor in the test system has its own controller. This controller uses the data from the torque sensor in the test setup as a feedback signal and controls the load motor to apply the desired load.

The test system uses one interface, which is not included in **Figure 7**. This interface sends commands to the ER-PMSM controller and the load motor controller simultaneously. In this way, two controllers work simultaneously.

Finally, the tests are carried out without the transmission mechanism. The ER-PMSM output shaft is directly connected to the load motor output shaft with the couplings shown in **Figure 7**. There are only couplings and torque sensor between load motor and ER-PMSM.

The parameters of the ER-PMSM are used in the experimental setup are given in **Table 1**. MPC parameters are selected as in **Table 2**. PID parameters are given in **Table 3**.

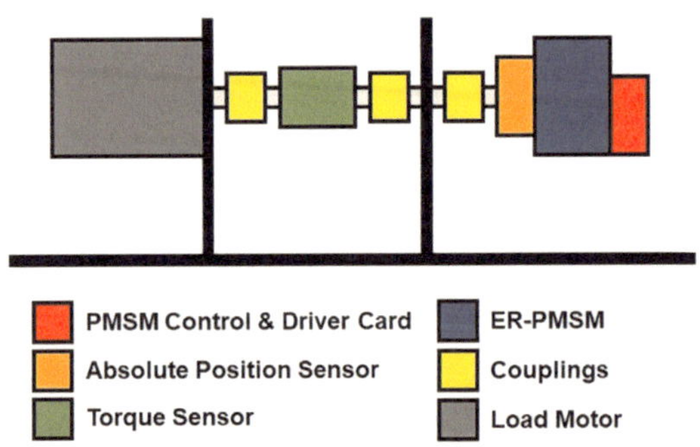

Figure 7.
Test system block diagram.

Parameter	Value
R_d, and R_q	$91\ m\Omega$
L_d	$57\ \mu H$
L_q	$78\ \mu H$
λ_{pm}	$2.9\ mWb$
p	21
J	$1.33 \cdot 10^{-4}\ kg \cdot m^2$

Table 1.
ER-PMSM parameters.

n_y	n_u	W_y	W_u	T_s
21	1	1	1	1 ms

Table 2.
MPC parameters.

Position Controller			Velocity Controller			Current Controller		
K_p	K_i	K_d	K_p	K_i	K_d	K_p	K_i	K_d
73	0	0	0.375	39	0	0.5	1250	0

Table 3.
PID parameters.

D-axis and q-axis current controllers are selected the same. The sampling time of the current controllers is determined as $50\mu s$. The position and speed control sampling times are set to 1 ms.

Experimental study is carried out in four different test condition. In these tests, both the sensitivity of position control is observed and the two controllers are compared with each other.

In the first test, position control sensitivity of ER-PMSMs under cogging torque is investigated. In the second and third tests, the performances of both controllers are compared. The performance comparison is made for the condition where there is no external disturbance factor. In the last test, the disturbance caused by external factors such load torque is examined.

3.1 Low speed position control test

In this test, the position reference profile is shown in **Figure 8**, which changes at a speed of 0.5 rad/s, is applied as the position reference. The purpose of the test is to

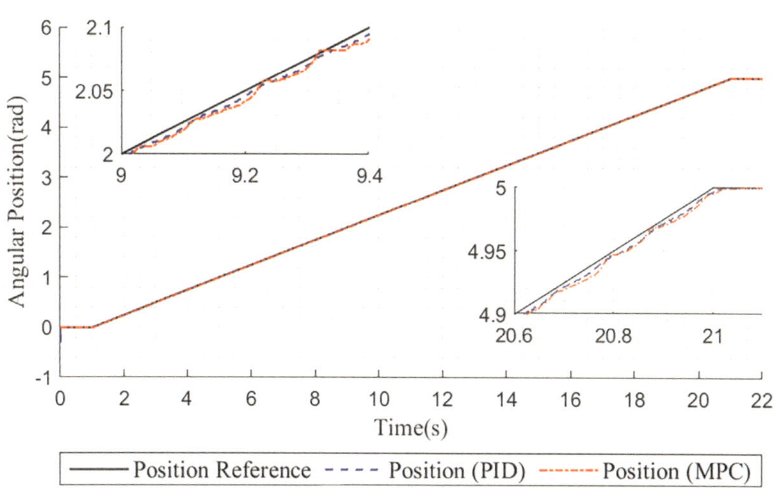

Figure 8.
Low speed position control test experimental results.

examine the effect of cogging torque on position control as the rotor position changes at a slow rate. In addition, there is no external load in this test.

In **Figure 8**, there are position control outputs for both controllers. It can be observed from **Figure 8** that, there are small ripples in position tracking in both controllers. These ripples are at 4 *mrad* in PID controller, as they are 10 *mrad* in MPC.

3.2 Sinusoidal reference position control test

In this test, a sinusoidal signal with an amplitude of 1 rad and a frequency of 15 Hz is set as the position reference. The purpose of this test is to examine the bandwidths of the controllers. In addition, there is no external load in this test.

In **Figure 9**, there are position control outputs for both controllers. As can be observed from **Figure 9**, the amplitude bandwidth of both controllers is same. However, phase shift in position control with MPC is higher than with PID.

Moreover, **Figure 9** shows the power consumption of both controllers during the sinusoidal motion. As can be seen from the power graph, control with MPC consumes less power than PID.

3.3 Step reference position control test

In this test, a step signal with an amplitude of 5 *rad* and a rate of 80 *rad/s* is set as the position reference. The purpose of this test is to observe the step response of both controllers. There is also no external load in this test.

In **Figure 10**, there are position control outputs of the both controllers are given. It can be observed from **Figure 10** that, rise time and settling time values are the same

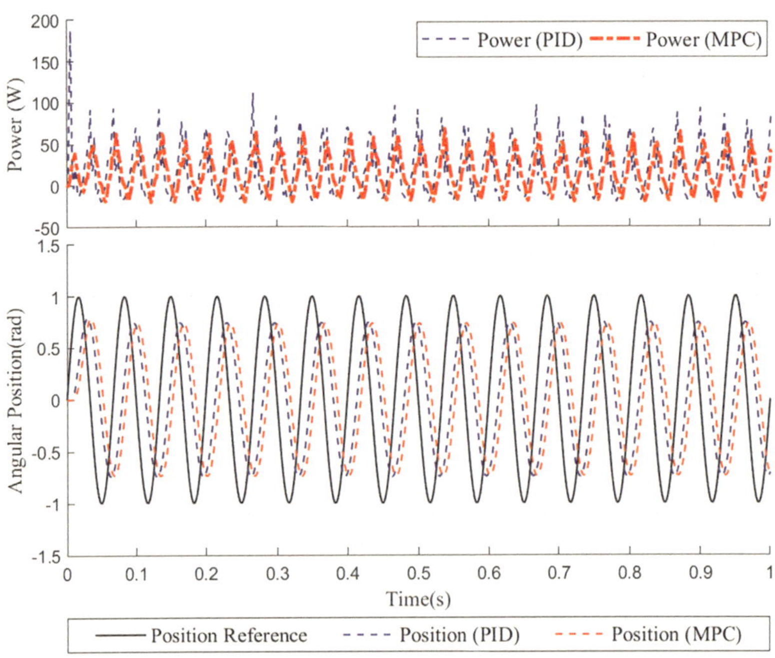

Figure 9.
Sinusoidal reference position control test experiment results.

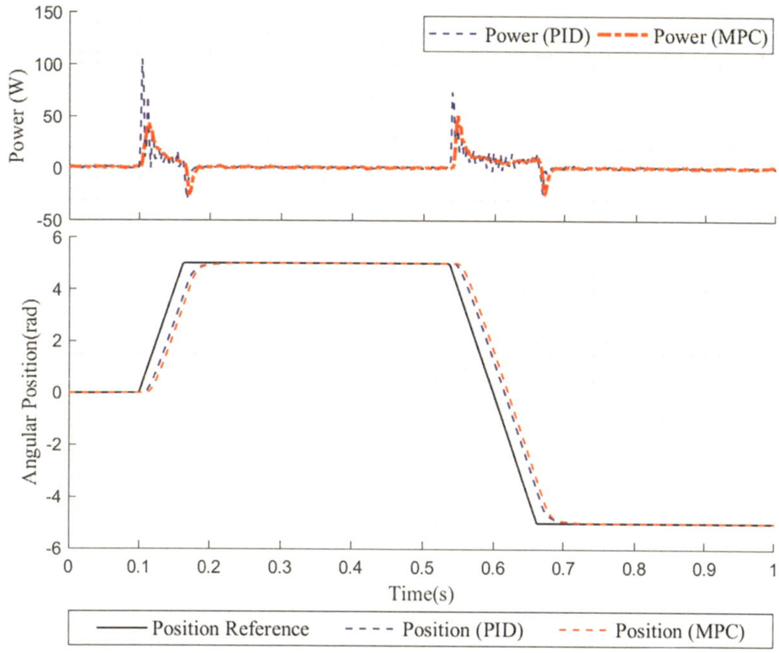

Figure 10.
Step reference position control test experiment results.

for both controllers. In MPC, the acceleration at the beginning of the motion is lower than in PID. However, the MPC slackened faster in the last part of the movement, both controllers completed the movement in the same time.

Also, **Figure 10** shows the power consumption of both controllers while making step motion. As can be seen from the power graph, control with MPC consumes less power than PID.

3.4 Low speed position control test with external load

In this test, the command in "Low-Speed Position Control Test" is used as the position reference. However, unlike the "Low-Speed Position Control Test", the external load in **Figure 11** is applied to the PMSM rotor shaft. This test investigates the effects of an external load, a.k.a. disturbance rejection performance.

As can be observed in **Figure 11**, position tracking where the load is constant is the same as in the "Low-Speed Position Control Test". When the load changes suddenly, the position control error increases up to 20 *mrad* for both controllers. MPC compensates for this error within 30 *ms*, while PID controller compensates for this error within approximately 90 *ms*.

4. Discussions

Four different tests are carried out both not only to observe the control performance of the ER-PMSM but also to compare the two designed controllers with each other.

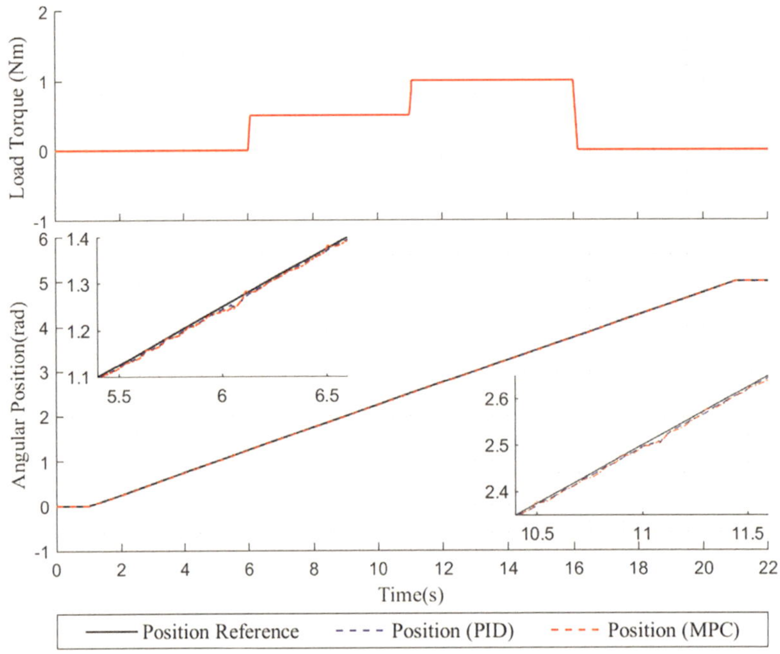

Figure 11.
Low speed position control test with external load experiment result.

As a result of "Low-Speed Position Control Tests" for both externally loaded and unloaded conditions, it has been observed that ER-PMSMs can be used in high-precision motion control applications. Cogging torque of ER-PMSMs causes ripples during movement. However, the level of these ripples is small and can be ignored in many applications. In addition, external loads on the rotor shaft do not have a negative effect on the position control of ER-PMSMs. Controllers can compensate these external loads rapidly.

After observing that the position control of ER-PMSMs can be done precisely with two different controllers, the advantages and disadvantages of these two controllers are investigated. First of all, the parameters of these two controllers are set to give similar performance. When the position outputs in the test results are examined, it is observed that the two controllers perform similarly.

It has been observed that the peak power consumption is lower in the control with MPC. The main reason for this behavior is that the acceleration characteristics of the two controllers are different. Although acceleration takes longer in MPC then PID, so the maximum acceleration level in MPC is lower then PID. In PID, on the other hand, the initial value of the acceleration is high, the acceleration time is shorter. Therefore, both controllers have similar performance in both sinusoidal reference and step signal reference, there are differences in power consumption.

Another difference between the two controllers is the processing loads. In the control with PID, it is necessary to control both at higher sampling frequencies and the much more operations. In MPC, on the other hand, fewer operations are performed with a low sampling frequency. Moreover, position measurement is sufficient for MPC, while current measurement is necessary for sequential PID control.

5. Conclusions

In this chapter, the use of ER-PMSMs in applications that require precise position control is investigated. Additionally, MPC is determined as the control method. PMSM position control structure is established using the CARIMA model. To compare the MPC performance, the sequential PID controller structure is established. These control methods are tested under four different test scenarios.

When the test results are examined, it is observed that the effect of cogging on control performance is relatively low even in a system without transmission elements (such as gearbox etc.) at the output shaft. Together with the transmission mechanism, the fluctuations caused by the cogging will be negligible. According to these results, it is shown that actuators with low transmission rates can be designed with ER-PMSMs.

In addition, when the test results of the two different control structures are examined, it is observed that the MPC is more suitable than PID controller in terms of energy consumption. It is also observed that the processing load of the MPC is less than PID controller and it can operate at a lower sampling time compared to the PID controller. As a result of this evaluation, for energy limited applications, it should be considered that would be more appropriate to use MPC in the control of ER-PMSMs.

Acknowledgements

Authors gratefully acknowledge reviewers' comments and the financial support of Roketsan Inc.

Author details

Goksel Kizir* and Abdul Furkan Kanburoglu
Roketsan Inc., Ankara, Turkey

*Address all correspondence to: goksel.kizir@roketsan.com.tr

References

[1] Mohanraj D, Gopalakrishnan J, Chokkalingam B, Mihet-Popa L. Critical aspects of electric motor drive controllers and mitigation of torque ripple—Review. IEEE Access. 2022;**10**: 73635-73674. DOI: 10.1109/ACCESS.2022.3187515

[2] Wang G, Valla M, Solsona J. Position Sensorless permanent magnet synchronous machine drives—A review. IEEE Transactions on Industrial Electronics. 2020;**67**(7):5830-5842. DOI: 10.1109/TIE.2019.2955409

[3] Xu D, Wang B, Zhang G, Wang G, Yu Y. A review of sensorless control methods for AC motor drives. CES Transactions on Electrical Machines and Systems. 2018;**2**(1):104-115. DOI: 10.23919/TEMS.2018.8326456

[4] Sayed E, Yang Y, Bilgin B, Bakr MH, Emadi A. A comprehensive review of flux barriers in interior permanent magnet synchronous machines. IEEE Access. 2019;**7**:149168-149181. DOI: 10.1109/ACCESS.2019.2947047

[5] De Klerk ML, Saha AK. A comprehensive review of advanced traction motor control techniques suitable for electric vehicle applications. IEEE Access. 2021;**9**:125080-125108. DOI: 10.1109/ACCESS.2021.3110736

[6] Zhu W, Chen D, Du H, Wang X. Position control for permanent magnet synchronous motor based on neural network and terminal sliding mode control. Transactions of the Institute of Measurement and Control. 2020;**42**(9): 1632-1640. DOI: 10.1177/0142331219893799

[7] Chen DB, Du HB, Jin XZ. Position tracking control for permanent magnet synchronous motor based on integral high-order terminal sliding mode control. In: Proceedings of 32nd Youth Academic Annual Conference of Chinese Association of Automation, 19–21 May 2017. Hefei, China: IEEE; 2017. pp. 234-239

[8] Cheng Y, Zhang J, Du H, Wen G, Lin X. Global event-triggered output feedback stabilization of a class of nonlinear systems. IEEE Transactions on Systems, Man, and Cybernetics. 2021; **51**(7):4040-4047. DOI: 10.1109/TSMC.2019.2931497

[9] Sahin M, Kizir G. Designing of Cascade controller by using the bees algorithm. International Journal of Control Systems and Robotics. 2022;**7**: 36-46

[10] Kiselev A, Kuznietsov A, Leidhold R. Position control of a permanent magnet synchronous motor using generalized predictive control algorithm. In: 2016 International Symposium on Power Electronics, Electrical Drives, Automation and Motion (SPEEDAM); 22–24 June 2016. Italy: IEEE; 2016. pp. 110-115

[11] Sahin M. Optimization of model predictive control weights for control of permanent magnet synchronous motor by using the multi objective bees algorithm. In: Hamid UZA, Faudzi AM, editors. Model-Based Control Engineering - Recent Design and Implementations for Varied Applications [Internet]. London: IntechOpen; 2021. DOI: 10.5772/intechopen.98810

[12] Zhao J, Ma T, Liu X, Zhao G, Dong N. Performance analysis of a coreless axial-flux PMSM by an improved magnetic equivalent circuit model. IEEE Transactions on Energy

Conversion. 2021;**36**(3):2120-2130. DOI: 10.1109/TEC.2020.3040009

[13] Miguel Guedes Rebelo J, Ângelo Rodrigues Silvestre M. Development of a modular controller to minimize current ripple in low inductance coreless permanent magnet motor. KEG [Internet]. 2020;**5**(6):757-768. Available from: https://knepublishing.com/index.php/KnE-Engineering/article/view/7095

[14] Habib A. A systematic review on current research and developments on coreless axial-flux permanent-magnet machines. IET Electric Power Applications. 2022;**16**(10):1095-1116. DOI: 10.1049/elp2.12218

[15] Tong W, Li S, Pan X, Wu S, Tang R. Analytical model for cogging torque calculation in surface-mounted permanent magnet motors with rotor eccentricity and magnet defects. IEEE Transactions on Energy Conversion. 2020;**35**(4):2191-2200. DOI: 10.1109/TEC.2020.2995902

[16] Wang K, Liang Y, Wang D, Wang C. Cogging torque reduction by eccentric structure of teeth in external rotor permanent magnet synchronous motors. IET Electric Power Applications. 2018; **13**(1):57-63. DOI: 10.1049/iet-epa.2018.5048

[17] Ahn JM, Son JC, Lim DK. Optimal Design of Outer-Rotor Surface Mounted Permanent Magnet Synchronous Motor for cogging torque reduction using territory particle swarm optimization. Journal of Electrical Engineering & Technology. 2021;**16**:429-443. DOI: 10.1007/s42835-020-00599-z

[18] Deng W, Zuo S, Chen W, Qian Z, Qian C, Cao W. Comparison of eccentricity impact on electromagnetic forces in internal- and external-rotor permanent magnet synchronous motors. IEEE Transactions on Transportation Electrification. 2022;**8**(1):1242-1254. DOI: 10.1109/TTE.2021.3112521

[19] Balashanmugham A, Maheswaran M. Permanent-magnet synchronous machine drives. In: El-Shahat A, Ruba M, editors. Applied Electromechanical Devices and Machines for Electric Mobility Solutions [Internet]. London: IntechOpen; 2019. Available from: https://www.intechopen.com/chapters/68961. DOI: 10.5772/intechopen.88597

[20] Secer G, Kizir G, Sahin M. Parameter estimation of permanent magnet synchronous machines based on a new model considering discretization effects of digital controllers. In: 2019 IEEE Applied Power Electronics Conference and Exposition (APEC);17–21 March 2019. USA: IEEE; 2019. pp. 1046-1051

[21] Rossiter JA. Model-Based Predictive Control a Practical Approach. 1st ed. Florida: CRC Press; 2004. p. 344. DOI: 10.1201/9781315272610

Chapter 3

Stereo-Measurement of 3D Poses of a Walking Person with a Puppet Model and Low-Rank Filter

Hiroshi Hanaizumi

Abstract

In order to achieve safety and security in public places by recognizing people's behavior, we have to measure a person's pose and capture his or her movement. Conventionally, it was hard to recognize the person itself before capturing the pose, and research has been carried out based on the silhouette of the person obtained by performing such as background subtraction. Efficient recognition was difficult in using silhouettes because they were greatly affected by the observation direction. Since the difficulty disappears if we adopt three-dimensional processing, some methods using Kinect have been proposed. Unfortunately, their performance was limited in the usage environment. Here OpenPose is applied to the stereo-pair images for getting the two-dimensional coordinates of the joint points. The coordinates may fluctuate depending on the lighting. Assuming the fluctuation follows a Gaussian distribution, the three-dimensional coordinates will be mixed with a larger noise component following a Cauchy distribution. Since the Cauchy distribution does not have a defined mean and variance, we can not achieve the desired smoothing even with averaging. The method described here provides stable poses that can withstand subsequent use by applying a low-rank filter and a puppet model in which the distance between joint points is invariant.

Keywords: stereo-camera, OpenPose, joint position, noise reduction, Cauchy distribution

1. Introduction

Recognizing the behaviors of people in public spaces such as train stations is very important to ensure our safety and security. For example, even if a person feels sick, falls, or has trouble with another person while walking, we can automatically recognize it and can help the person immediately. A person's behavior can be defined as a time trace pattern of poses of the person. In other words, if the pose can be quantified, it is possible to recognize the behavior as a pattern of temporal change. For the quantification, it is first essential to be able to measure the pose with high precision. There have been many methods for recognizing human poses [1–3]. They can be divided into two groups: silhouette-based methods and joint-based ones. In the former group, human silhouettes were obtained, for example, by a background subtraction or by calculating optical flow [4], and were used to analyze the behavior. In the latter, joint positions measured by Kinect were used [5–8]. The silhouette-based methods, however, had difficulties in estimating the poses. It was not easy to find some feature points to describe the pose from the silhouette. On the other hand, the joint-based methods easily described the pose as the placing of the joint points. In particular, Kinect gave us three-dimensional positions of them and made the analysis easier. Unfortunately, Kinect projected a dot pattern to obtain the distance to a target, so there were some problems such as limitations in measurable distance and outdoor environment. OpenPose [9] was also one of the solutions. It gave us 25 keypoints, including human joints on an image. The two-dimensional coordinates were useful in describing poses in a walk perpendicularly crossing the optical axis [10]. The OpenPose-based method had similar problems to the silhouette-based one. The size change of the pose with the distance made the analysis more complicated. Some solutions have been required.

In this paper, we propose a measurement method for obtaining three-dimensional joint coordinates of a human pose by applying OpenPose to a sequence of stereo-pair images. We obtain the three-dimensional joint coordinates from the two-dimensional coordinate pair of the same joint extracted by OpenPose in each of the stereo-pair images. This means that we quantify the pose of a person in the three-dimensional space with a small number of joints. Recently, OpenPose provided the three-dimensional keypoint features with a demo. It, however, requires a specific multi-camera system and has some barriers to use for general motion measurements. Here we construct a software system for obtaining three-dimensional joint coordinates using a single general-purpose stereo camera. The stereo-measurement also has the feature of being able to acquire a person's pose from a distance. Therefore, we can measure the poses of multiple persons simultaneously in a wide area. The joint coordinates reconstructed from stereo-pair two-dimensional ones are severely affected by missing values and/or some fluctuations contained in OpenPose output. Therefore, noise reduction is essential in the proposed method. When we obtain the three-dimensional coordinates, a parallax on stereo-pair images is included in the denominator of the stereo-vision formula. Therefore, Gaussian fluctuation in the stereo-pair images appears as a larger noise component following a Cauchy distribution on the three-dimensional coordinates. Since the mean and the variance of the Cauchy distribution are undefined, simple averaging against Gaussian noise does not achieve an expected smoothing. We need another approach for the noise reduction in the three-dimensional coordinates. We regard walking as being consisted of a kind of periodical movement with a spatial correlation structure among all joints. Therefore, we focus on the dominant mode of them as the signal component to be measured. We realize the noise reduction by extracting the low-rank component [11] of the time sequence of joint coordinates recognized by OpenPose. We call It a low-rank filter (LRF). Applying LRF separately to each joint data may destroy the spatial structure. Therefore, we compensate the joint coordinates using a puppet model [12] introduced from the fact that every bone length, that is distance between two joints, is invariant for any pose. Severe noise included in measured coordinates of the joints yields many candidates of the pose to be obtained. Those poses include meaningless ones as the human posture. The puppet model enables us to eliminate the meaningless poses by fixing the length of the upper and lower limbs without restricting the angle between the limbs. The length fixing acts as some kind of constraint. Thus, knowledge about the object makes meaningful measurement and recognition even in adverse conditions. In the following sections, we describe the principle and the procedure of the method. Application results of an actual person walking are also shown.

2. Principle

In order to realize direction-free analysis for human walking, we consider a remotely joint-based measurement using stereo-vision. In the conversion from the joint coordinates in both stereo-pair images to three-dimensional ones, Gaussian noise expands to a terrible one following the Cauchy distribution. Therefore, the main task of this research is to reduce the severe noise and to get accurate three-dimensional joint coordinates. We consider that human joints do not move randomly but move with a mutual spatial correlation structure. From this fact, we introduce a puppet model that acts to restrict the joint position in an area.

2.1 World and camera coordinate system

Figure 1 shows world and camera coordinate systems used here. A set of stereo-camera C_1 and C_2 are calibrated (distortion correction and collimation are performed) so that their optical axes are parallel and have the same focal length with distance B. They are installed so that the center of the camera C_1 coincides with the origin of the world coordinate system.

A point $P(x, y, z)$ on the world coordinate system is observed as image points $p_1(u_l, v_l)$ and $p_2(u_r.v_r)$ on cameras C_1 and C_2, respectively. The coordinates x, y, z are obtained as

$$x = \frac{u_l B}{u_l - u_r}, y = \frac{v_l B}{u_l - u_r}, z = \frac{fB}{u_l - u_r}, \tag{1}$$

where, f is the focal length and B baseline length. The camera coordinates u_l, v_l and u_r, v_r are also calculated from the coordinates x, y, z as

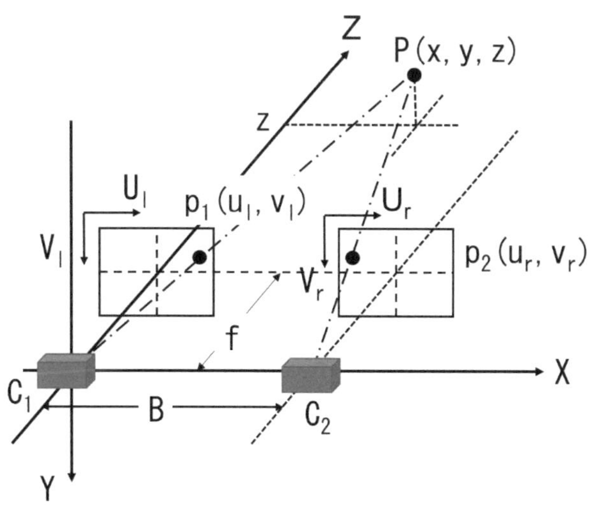

Figure 1.
World and camera coordinate systems.

$$u_l = f\frac{x}{z}, u_r = f\frac{x-B}{z}, v_l = v_r = f\frac{y}{z}. \quad (2)$$

Gaussian noise contained in joint coordinates u_l, v_l, u_r, and v_r observed on both stereo-pair images appear in each of the numerator and the denominator of Eq.(1). Since the ratio of two random normal distributed variables with mean 0 and different variances follows the Cauchy distribution [13], obtained three-dimensional joint coordinates x, y, and z are severely affected by noise following a Cauchy distribution. The first and second moments of the Cauchy distribution diverge to infinity, its mean and variance are not defined. This means a simple averaging does not achieve the expected smoothing.

2.2 OpenPose and keypoint coordinates

OpenPose is the first real-time multi-person system to jointly detect human body, hand, facial, and foot keypoints [9]. Here we use it as a human joint detector and adopt 15 keypoints for human walking analysis among the output format BODY_25, as shown in **Figure 2**. OpenPose provides the keypoint with detection confidence in a JSON file. The confidence takes a value between 0 and 1, and the higher the value the more confident it is. When a joint is hidden by the body, its confidence takes 0, and the coordinates are also 0. Therefore, the missing coordinates such as due to joint occlusion yield larger noise. We have to complete the missing values first. OpenPose may also cause incorrect detections, such as leg swapping, arm swapping, leg duplication, and arm duplication, due to poor lighting conditions in image capturing. In those cases, the confidence takes lower but non-zero values, so it is hard to detect them using confidence. Both the swapping and the duplication also cause larger noise, so we need some algorithms to detect and correct them.

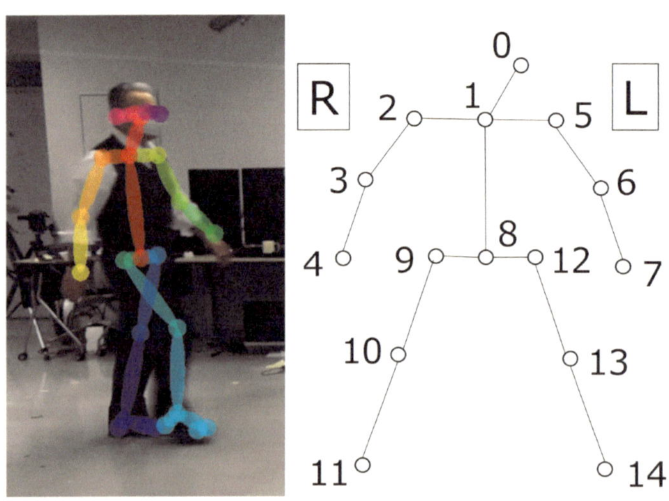

(a) *Actual keypoints detected.* (b) *Keypoints used.*

Figure 2.
OpenPose keypoints in BODY_25.

2.2.1 Missing value interpolation

Our strategy for interpolating the missing values is to fill them in a simple way, then smooth them with preserving features. Let $\{s_1, \dots, s_i, \dots, s_j, \dots, s_n\}$ be a sequence of measured coordinates with some missing values as

$$s_k \begin{cases} = 0 & (i \le k \le j) \\ \ne 0 & (others) \end{cases}. \tag{3}$$

Those missing values are interpolated by using internal division points of both ends of non-zero-confidence observations as

$$s_k = \frac{s_{i-1}(j-k+1) + s_{j+1}(k-i+1)}{j-i+2} \qquad (i \le k \le j). \tag{4}$$

When $j = n$, we use $s_k = s_{i-1} \ (i \le k \le n)$, and $i = 1$ we use $s_k = s_{j+1} \ (1 \le k \le j)$.

2.2.2 Keypoint swapping and duplication

In wrong recognitions, left and right legs are duplicated as one leg, and leg swapping also occurs more frequently. In normal walking, one of the legs is swung to the front and fixed on the floor, and the other is also swung and fixed alternately. The left and right ankles draw continuous trajectories that accurately reflect these movements. When the leg swapping and/or duplication occur, some discontinuities are yielded in the trajectories. Therefore, we detect them by finding the discontinuities. We detect the swapping by finding the discontinuity at the same position on both left and right ankle trajectories, and the duplication by finding it on one of them. For simplicity, consider a person walking parallel to the x-axis as shown in **Figure 2**. Let x-coordinate vectors of left and right ankles be $\boldsymbol{s}_L = (s_{L1} s_{L2} \dots s_{Ln})^T$ and $\boldsymbol{s}_R = (s_{R1} s_{R2} \dots s_{Rn})^T$ for n observations. Their elements s_{Li} and $s_{Ri} (i = 1, \dots, n)$ are relative position to a keypoint mid-hip. The swapping is detected at the i-th element when $\delta_{Li}/w > threshold$ and $\delta_{Ri}/w > threshold$, where, w means the stride length of the walk as

$$w = \min(\max(\boldsymbol{s}_L) - \min(\boldsymbol{s}_L), \max(\boldsymbol{s}_R) - \min(\boldsymbol{s}_R)), \tag{5}$$

$$\delta_{Li} = |s_{Li} - s_{Li-1}|, \tag{6}$$

$$\delta_{Ri} = |s_{Ri} - s_{Ri-1}|. \tag{7}$$

2.2.3 Vertical averaging

As already mentioned, stereo-cameras are also parallelized by the calibration. Therefore, there should be no difference in vertical coordinates between corresponding joints on the stereo-pair images. If the difference exists, it is considered to be due to noise, and we replace those vertical coordinates with their average values. Thus, OpenPose joint data are completed.

2.3 Pose vector and behavior matrix

In this study, we define behavior as the temporal change of the poses. Therefore, it is necessary to quantify each pose for describing the behavior. We define keypoint vector $\boldsymbol{q}_t$ and the pose vector $\boldsymbol{p}_t$ independently of the dimension. The former is described as

$$\boldsymbol{q}_t = \left(u_0^t \; v_0^t \; u_1^t \; v_1^t \; \cdots \; u_{m-1}^t \; v_{m-1}^t\right), \tag{8}$$

on an image or

$$\boldsymbol{q}_t = \left(x_0^t \quad y_0^t \quad z_0^t \quad x_1^t \quad y_1^t \quad z_1^t \quad \cdots \quad x_{m-1}^t \quad y_{m-1}^t \quad z_{m-1}^t\right), \tag{9}$$

in the world coordinate system, where, (u_i, v_i) and (x_i, y_i, z_i) $(i = 0, \cdots, m-1)$ are two- and three-dimensional coordinates of the keypoints, respectively, and $m = 15$. Of course, since the 2D pose strongly depends on the direction of observation, the 3D pose vector without such dependence is valid. Therefore, when we simply refer to the pose vector, we mean the 3-D one. Using their relative coordinates, the latter is represented as

$$\boldsymbol{p}_t = \left(u_0^t - u_8^t \quad v_0^t - v_8^t \quad \cdots \quad \cdots \quad u_{m-1}^t - u_8^t \quad v_{m-1}^t - v_8^t\right), \tag{10}$$

on an image or

$$\boldsymbol{p}_t = \left(x_0^t - x_8^t \quad y_0^t - y_8^t \quad z_0^t - z_8^t \quad \cdots \quad x_{m-1}^t - x_8^t \quad y_{m-1}^t - y_8^t \quad z_{m-1}^t - z_8^t\right), \tag{11}$$

in the world coordinate system, where, (u_8, v_8) and (x_8, y_8, z_8) are coordinates of the keypoint mid-hip at time t, and $\boldsymbol{p}_t$ does not include it while $\boldsymbol{q}_t$ does. We consider that the keypoint mid-hip is located near the true centroid of a human body, so we use it instead of the centroid. Thus, the pose vector is useful to represent relative pose regardless of the position. Finally, we define keypoint matrix Q_n and behavior matrix P_n for n observations as

$$Q_n = \begin{pmatrix} \boldsymbol{p}_0 \\ \boldsymbol{p}_1 \\ \vdots \\ \boldsymbol{p}_{n-1} \end{pmatrix}, P_n = \begin{pmatrix} \boldsymbol{p}_0 - \boldsymbol{\mu} \\ \boldsymbol{p}_1 - \boldsymbol{\mu} \\ \vdots \\ \boldsymbol{p}_{n-1} - \boldsymbol{\mu} \end{pmatrix}, \boldsymbol{\mu} = \frac{1}{n}\sum_{i=0}^{n-1} \boldsymbol{p}_i. \tag{12}$$

The covariance matrix of the behavior matrix is also obtained as

$$\Sigma = \frac{1}{n-1} P_n^T P_n. \tag{13}$$

Since the elements within the covariance matrix represent the degree of correlation between corresponding coordinates, we consider that it includes important information about the spatial correlation structure of joint movements.

2.4 Auto-regression model and low-rank filter

OpenPose may yield some fluctuations in its recognitions, and pose vectors still have noise due to the fluctuation. Since they severely affect the three-dimensional coordinates, we have to reduce the noise in the pose vectors. On the other hand, the coordinates have some correlation structures, such that they do not change abruptly in time while keeping their relative positions. Therefore, we realize the noise reduction by extracting low-rank components from the observed time sequential joint coordinates. The one-dimensional coordinate sequence is once converted to a two-dimensional matrix shape to extract the low-rank components. After we apply a rank reduction operation to that matrix, the resultant matrix is converted back to the original one-dimensional sequence. To perform the process, we need an interconversion algorithm between a one-dimensional signal and a two-dimensional matrix. Let $\boldsymbol{U}_k$ be a temporal observation vector of positions of the k-th joint of a moving person on the image as

$$\boldsymbol{U}_k = \begin{pmatrix} u_k^0 & u_k^1 & \cdots & u_k^n \end{pmatrix}, \tag{14}$$

where, the superscripts show time, that is, frame. Since the movement is not so fast and not random, they have some correlation with each other. This means that they fit into an auto-regressive (AR) model with length of m as

$$\begin{cases} u_k^m = a_0 u_k^0 + a_1 u_k^1 + \cdots + a_{m-1} u_k^{m-1} \\ u_k^{m+1} = a_0 u_k^1 + a_1 u_k^2 + \cdots + a_{m-1} u_k^m \\ \vdots \\ u_k^n = a_0 u_k^{n-m} + a_1 u_k^{n-m+1} + \cdots + a_{m-1} u_k^{n-1} \end{cases} . \tag{15}$$

These equations show that the current value of a one-dimensional signal is determined by the values of the m signals preceding it. Using matrix format, we rewrite Eq. (15) as

$$\begin{pmatrix} u_k^0 & u_k^1 & \cdots & u_k^{m-1} \\ u_k^1 & u_k^2 & \cdots & u_k^m \\ \vdots & \vdots & \ddots & \vdots \\ u_k^{n-m} & u_k^{n-m+1} & \cdots & u_k^{n-1} \end{pmatrix} \begin{pmatrix} a_0 \\ a_1 \\ \vdots \\ a_{m-1} \end{pmatrix} = \begin{pmatrix} u_k^m \\ u_k^{m+1} \\ \vdots \\ u_k^n \end{pmatrix}, \tag{16}$$

and for short as

$$X\boldsymbol{A} = \boldsymbol{Y}, \tag{17}$$

where matrix X is called the Hankel matrix. The vector $\boldsymbol{Y}$ is calculated from the matrix X, and also X is reconstructed from $\boldsymbol{Y}$. Considering that the observations consist of dominant signal component with some correlation structure and some noise, Eq. (17) can be described as

$$(X_s + X_n)\boldsymbol{A} = (\boldsymbol{Y}_s + \boldsymbol{Y}_n). \tag{18}$$

Here we realize the low-rank filter for the vector $\boldsymbol{Y}_s$ by separating X_s with rank r from X_n as

$$X_s \boldsymbol{A}_s = \boldsymbol{Y}_s \quad (rank X_s = r_f). \tag{19}$$

Since the coefficient vector $\boldsymbol{A}$ in Eq. (19) is different from that in Eq. (17), an iterative algorithm is needed for the calculation [11].

2.5 Spatial AR model and filtering

As we have already defined, human behavior is a multi-temporal set of poses. We have focused on the temporal correlation structure of each joint in human behavior, and constructed the low-rank filter. Here as an analogy from temporal correlation to spatial one, we propose a spatial AR model. The model is based on the fact that we can estimate the pose of the whole body even if some joints are hidden and cannot be observed. We consider that the linear form of the matrix X in Eq. (17) and its reproducibility from the coordinate vector $\boldsymbol{Y}$ are essential in the low-rank filtering. Therefore, we introduce a matrix and a coordinate vector in the same manner. Let $\boldsymbol{x}_j^t$ be the j-th joint coordinates at time t, $\mathbf{0}$ zero coordinates, and $\boldsymbol{a}_k$ the k-th coefficient vector as

$$\boldsymbol{x}_j^t = \begin{pmatrix} x_j^t & y_j^t & z_j^t \end{pmatrix}, \mathbf{0} = \begin{pmatrix} 0 & 0 & 0 \end{pmatrix}, \boldsymbol{a}_k = \begin{pmatrix} a_k^x & a_k^y & a_k^z \end{pmatrix}. \tag{20}$$

By analogy to the description of Eq.(16), we propose a new linear matrix equation as

$$\begin{pmatrix} \mathbf{0} & \boldsymbol{x}_1^{t-2} & \cdots & \boldsymbol{x}_{m-1}^{t-2} \\ \boldsymbol{x}_0^{t-2} & \mathbf{0} & \cdots & \boldsymbol{x}_{m-1}^{t-2} \\ \vdots & \vdots & \ddots & \vdots \\ \boldsymbol{x}_0^{t-2} & \boldsymbol{x}_1^{t-2} & \cdots & \mathbf{0} \\ \mathbf{0} & \boldsymbol{x}_1^{t-1} & \cdots & \boldsymbol{x}_{m-1}^{t-1} \\ \boldsymbol{x}_0^{t-1} & \mathbf{0} & \cdots & \boldsymbol{x}_{m-1}^{t-1} \\ \vdots & \vdots & \ddots & \vdots \\ \boldsymbol{x}_0^{t-1} & \boldsymbol{x}_1^{t-1} & \cdots & \mathbf{0} \\ \mathbf{0} & \boldsymbol{x}_1^{t} & \cdots & \boldsymbol{x}_{m-1}^{t} \\ \boldsymbol{x}_0^{t} & \mathbf{0} & \cdots & \boldsymbol{x}_{m-1}^{t} \\ \vdots & \vdots & \ddots & \vdots \\ \boldsymbol{x}_0^{t} & \boldsymbol{x}_1^{t} & \cdots & \mathbf{0} \end{pmatrix} \begin{pmatrix} a_0 \\ a_1 \\ \vdots \\ a_{3m-1} \end{pmatrix} = \begin{pmatrix} \boldsymbol{x}_0^{t-2} \\ \vdots \\ \boldsymbol{x}_{m-1}^{t-2} \\ \boldsymbol{x}_0^{t-1} \\ \vdots \\ \boldsymbol{x}_{m-1}^{t-1} \\ \boldsymbol{x}_0^{t} \\ \vdots \\ \boldsymbol{x}_{m-1}^{t} \end{pmatrix}. \tag{21}$$

The model is characterized by symmetrical calculation such that each joint coordinates x, y, and z on the right side are obtained from all non-zero coordinates on the left side. We believe the symmetrical calculation enables us to realize stable analysis independently of the walking direction. Since the number of unknowns increases threefold, we need at least sequential three poses for the calculation. We use more poses to get the smoothing effect. The filtering is achieved by decreasing the rank of the left-side matrix.

2.6 Puppet models

Eq. (1) tells us that fluctuations in observed two-dimensional coordinates $\boldsymbol{u}_l$ and $\boldsymbol{u}_r$ are expanded and yield severe noise in the three-dimensional ones. Since the noise reduction by using the low-rank filter above is independently applied to each joint observation, it is not enough to reconstruct the meaningful three-dimensional pose.

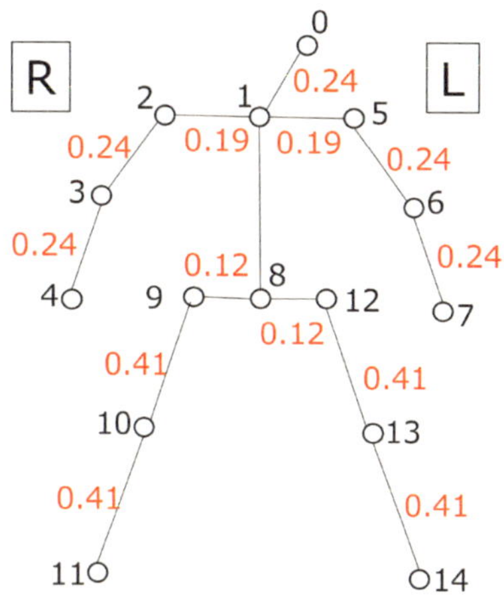

Figure 3.
Puppet model with joint number (black) and limb lengths (red) [m].

Some constraints are required. Here we propose two types of puppet models as pose stabilizers. The model is based on the fact that the distance between the neighboring joints is invariant of the pose. **Figure 3** shows the puppet with the joint number (black) and limb length in [m] (red). The puppet has a symmetrical structure with a predetermined length limb and freely moving joints in any direction. We determined their lengths based on roughly measuring the length of the limbs of a person with a height of 1.7 [m]. We introduce two types of models with different constraint methods: the flexible type and the rigid one. We call the former Puppet-I [12] and the latter Puppet-II. In both models, we regard the keypoint 8 (mid-hip) as the reference joint.

2.6.1 Puppet-I

In this model, we evaluate the limb length between a neighboring joint pair whose positions are calculated by using Eq. (1). If the length does not match that of the model, we regard that there is an error in the position of the joint u_r on the right camera image, and adjust the parallax $u_l - u_r$ so that they match. This process is started from the reference joint. Let the position of the i-th neighboring joint $P_i(x_i, y_i, z_i)$ be calculated from the stereo-pair image points $Q_i^l(u_i^l, v_i^l)$ and $Q_i^r(u_i^r, v_i^r)$, and $P_k(x_k, y_k, z_k)$ be the reference. The parallax adjustment is performed as

$$\Delta u_{opt}^r \leftarrow \arg\min_{\Delta u^r} \left| \sqrt{(x_i - x_k)^2 + (y_i - y_k)^2 + (z_i - z_k)^2} - L_{ik} \right|, \tag{22}$$

where,

$$x_i = \frac{B u_i^l}{u_i^l - u_i^r + \Delta u^r}, \tag{23}$$

$$y_i = \frac{B v_i^l}{u_i^l - u_i^r + \Delta u^r}, \tag{24}$$

$$z_i = \frac{Bf}{u_i^l - u_i^r + \Delta u^r}, \tag{25}$$

and L_{ik} is the predetermined distance from the reference to the i-th joint. The adjustment is repeatedly applied to successive joint pairs, such as mid-hip to hip, hip to knee, and knee to ankle for legs, and mid-hip to neck, neck to shoulder, shoulder to elbow, and elbow to wrist for arms. The optimization process in Eq. (22) is similar to the finding solution for a quadratic-like curve as shown in **Figure 4**. **Figure 4** indicates that we may have two solutions (left) and have nothing (right). In the former case, we select the best combination so that, for example, both shoulders have the longest distance. Arms and legs are also compensated so that they have the largest angle at the elbow and knee, respectively. In the latter case, the compensation is performed so that the difference becomes the minimum. Substituting the optimal adjuster Δu_{opt}^r for the i-th joint into Eqs. (23)–(25), we obtain the three-dimensional coordinate of the joints.

2.6.2 Puppet-II

In this model, we regard that relative positions among body joints, that is, mid hip, left hip, right hip, neck, left shoulder, and right shoulder, are fixed independently of poses. Assuming the rotation angles α, β, γ and the shift x_s, y_s, z_s, we transfer the model shown in **Figure 5a** into the world coordinate system. Then, we predict the position of their joints in both camera coordinate systems. We regard that the measurement was made when the squared sum of the difference among predicted positions

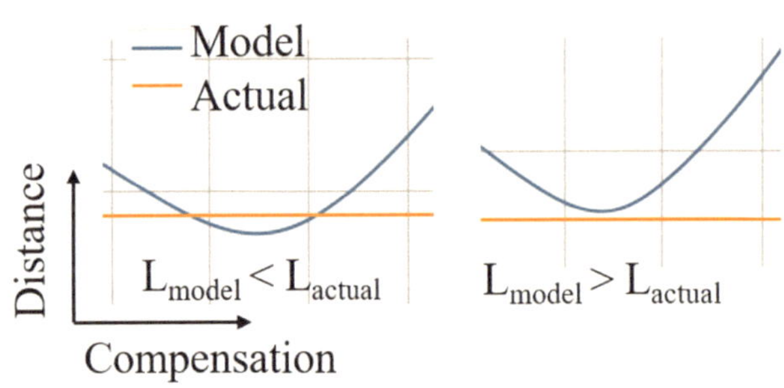

Figure 4.
Model limb length and solutions.

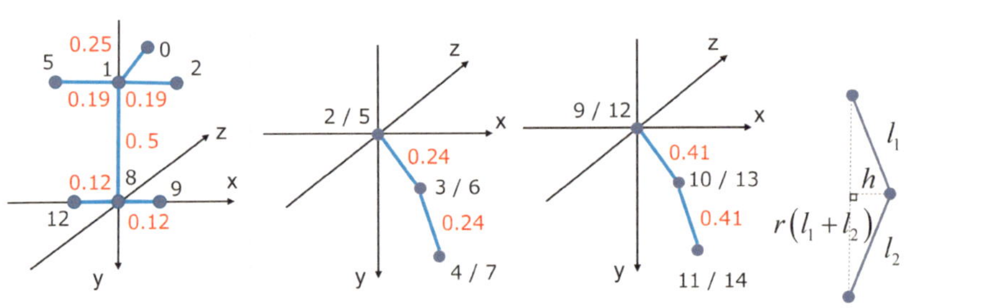

(a) *Body model.* (b) *Arm model.* (c) *Leg model.* (d) *Model triangle.*

Figure 5.
Rigid body model, upper limb model, and lower one with limb lengths.

(u_l^p, v_l^p), (u_r^p, v_r^p) and observed ones (u_l^o, v_l^o), (u_r^o, v_r^o) on the camera coordinates for six joints within the part *body* falls to the minimum and that we obtain the optimum angles and shifts as.

$$\mathbf{E}_{opt}^{body} \leftarrow \underset{\alpha,\beta,\gamma,x_s,y_s,z_s}{\arg\min} \sum_{j \in body} \left\{ \left(u_{l,j}^o - u_{l,j}^p\right)^2 + \left(v_{l,j}^o - v_{l,j}^p\right)^2 + \left(u_{r,j}^o - u_{r,j}^p\right)^2 + \left(v_{r,j}^o - v_{r,j}^p\right)^2 \right\}, \tag{26}$$

where, $\mathbf{E}_{opt}^{body}$ is the optimized angles and shifts as

$$\mathbf{E}_{opt}^{body} = \begin{pmatrix} \alpha_{opt} & \beta_{opt} & \gamma_{opt} & x_{opt} & y_{opt} & z_{opt} \end{pmatrix}, \tag{27}$$

and the summation means sum over six joints in the body model (nose is not used), and camera image coordinates are calculated from the initial position of body joints $\left(x_j^{init}, y_j^{init}, z_j^{init}\right)$ $(j \in body)$ as

$$\begin{pmatrix} x_j^p \\ y_j^p \\ z_j^p \end{pmatrix} = R(\alpha,\beta,\gamma) \begin{pmatrix} x_j^{init} \\ y_j^{init} \\ z_j^{init} \end{pmatrix} + \begin{pmatrix} x_s \\ y_s \\ z_s \end{pmatrix} \quad (j \in body), \tag{28}$$

$$R(\alpha,\beta,\gamma) = \begin{pmatrix} 1 & 0 & 0 \\ 0 & \cos\alpha & -\sin\alpha \\ 0 & \sin\alpha & \cos\alpha \end{pmatrix} \begin{pmatrix} \cos\beta & 0 & -\sin\beta \\ 0 & 1 & 0 \\ -\sin\beta & 0 & \cos\beta \end{pmatrix} \begin{pmatrix} \cos\gamma & -\sin\gamma & 0 \\ \sin\gamma & \cos\gamma & 0 \\ 0 & 0 & 1 \end{pmatrix}, \tag{29}$$

$$u_{l,j}^p = f\frac{x_j^p}{z_j^p}, \quad u_{r,j}^p = f\frac{\left(x_j^p - B\right)}{z_j^p}, \quad v_{l,j}^p = v_{r,j}^p = f\frac{y_j^p}{z_j^p}. \tag{30}$$

The three-dimensional coordinates of joints in the body are obtained by substituting the optimum angles and shifts into Eq. (28). After body joints measurement, those in the arms and legs are processed. As shown in **Figure 5b** and **c**, arm and leg models have the same structure. Therefore, we can apply the same process to both arm and leg joint measurements. We modify the processes described in Eq. (26)-(30) to, for example, that for left the arm *l_arm*, including joints 6 and 7 as

$$\mathbf{E}_{opt}^{l\_arm} \leftarrow \underset{\alpha,\beta,\gamma,r}{\arg\min} \sum_{j \in l\_arm} \left\{ \left(u_{l,j}^o - u_{l,j}^p\right)^2 + \left(v_{l,j}^o - v_{l,j}^p\right)^2 + \left(u_{r,j}^o - u_{r,j}^p\right)^2 + \left(v_{r,j}^o - v_{r,j}^p\right)^2 \right\}, \tag{31}$$

where, r is a bending coefficient $(0 \le r \le 1)$ between the upper limb l_1 and the lower one l_2 in the triangle shown in **Figure 5d**, and

$$h = \frac{2S}{r(l_1 + l_2)}. \tag{32}$$

The area S of the triangle is calculated using Heron's formula. After initial coordinates setting, the rotation matrix $R(\alpha,\beta,\gamma)$ is applied to them, then they are shifted to the corresponding body joint, that is, the left shoulder in this case as.

Joint	X	y	z
2	0	0	0
3	-h	$-\sqrt{l_1^2 - h^2}$	0
4	0	$-r(l_1 + l_2)$	0
	0	0	0
6	h	$-\sqrt{l_1^2 - h^2}$	0
7	0	$-r(l_1 + l_2)$	0
	0	0	0
10	-h	$-\sqrt{l_1^2 - h^2}$	0
11	0	$-r(l_1 + l_2)$	0
	0	0	0
13	h	$-\sqrt{l_1^2 - h^2}$	0
14	0	$-r(l_1 + l_2)$	0

Table 1.
Initial coordinates for arm and leg models.

$$\begin{pmatrix} x_j^p \\ y_j^p \\ z_j^p \end{pmatrix} = R(\alpha, \beta, \gamma) \begin{pmatrix} x_j^{init} \\ y_j^{init} \\ z_j^{init} \end{pmatrix} + \begin{pmatrix} x_5 \\ y_5 \\ z_5 \end{pmatrix} \quad (j \in l\_arm). \tag{33}$$

After the optimization processing, elements of the vector $\boldsymbol{E}_{opt}^{l\_arm}$ are substituted into Eq. (33) for obtaining the three-dimensional coordinates of joints 6 and 7, that is, left elbow and left wrist. Those of other parts, right arm, left leg, and right leg, are also determined using the same manner. **Table 1** shows the initial coordinates of joints in each part optimization. Thus, three-dimensional coordinates of all the joints are measured.

3. Procedures

The followings are the procedures of the proposed method. We measure the three-dimensional position of 15 joints in model Body_25. The main process is noise reduction in the two-dimensional joint coordinates recognized by OpenPose on stereo-pair images.

Step 1: Reading keypoint data from a JSON file and checking their consistencies.

The keypoint data for a person in the image may be divided into a few portions and be output as a few persons' data. Mis-recognitions such as ghost keypoints also appear due to the very high sensitivity. The portion collection is made by using keypoint consistency between the stereo-pair images.

Step 2: Interpolating the coordinates of missing keypoints.

Using the confidence in the JSON file, we determine whether there is a missing keypoint or not. Then, the missing values are interpolated.

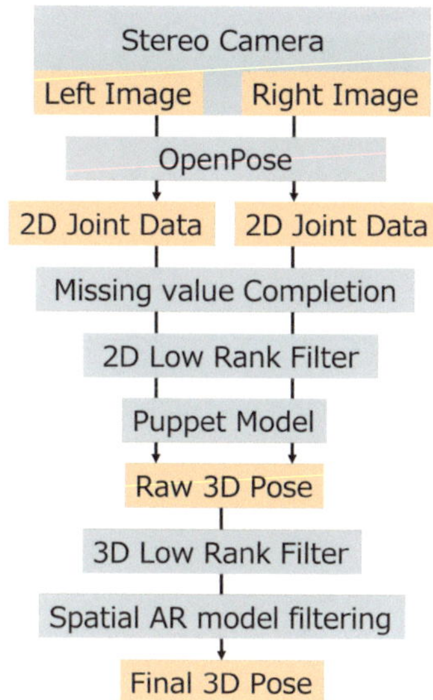

Figure 6.
Processing flow of the proposed method.

Step 3: Correcting the leg swapping and duplication.

They are detected by using the normalized speed of both ankles and are corrected. Arm swapping and duplication are also completed.

Step 4: Applying the low-rank filter (LRF-2D).

The LRF-2D is applied to the time sequence of each two-dimensional coordinate of each joint on both stereo-pair images.

Step 5: Constructing three-dimensional poses.

Three-dimensional poses are constructed by applying puppet-I or puppet-II.

Step 6: Applying LRF-3D.

LRF-3D is applied to the time sequence of each three-dimensional coordinate of each joint in the constructed poses.

Step 7: Applying the spatial AR model filtering (SARF).

SARF is applied to a three-dimensional pose in all frames.

Figure 6 shows the processing flow of the proposed method in which devices or functions are blue-backed and resultant data orange-backed.

4. Experiments and discussion

Here we show the stereo camera used and its calibration result, observed joint data, and some preprocessing results for them. Performances of the low-rank filter, spatial AR model, and puppet models are also discussed with the implementation.

4.1 Stereo camera and its calibration

Figure 7 shows the stereo camera we used in the experiments below, and **Table 2** shows its specification. The camera was stereo-calibrated so that parallelization was made. We evaluated the performance of the calibration by showing the difference in vertical coordinates between both the stereo-pair images of corner points of a chess

Figure 7.
A stereo camera Zed2 was used.

Parameter	Value
Number of pixels	3840 × 1080
Format	Side by side
Frames per second	30 fps
Baseline length	120 mm
Field of view	120 degrees

Table 2.
*Specification of the stereo camera. *Zed2 originally has higher performance. We used lower mode for easy recording.*

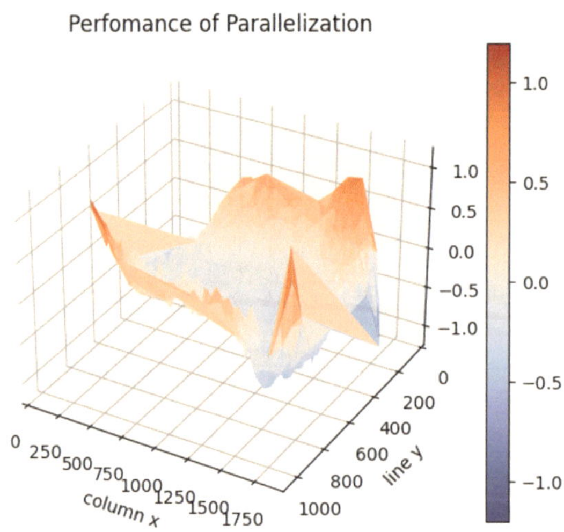

Figure 8.
Vertical coordinate differences among all corresponding points in stereo-pair images.

board sheet used for the camera calibration as shown in **Figure 8**. The difference was within plus or minus 0.5 pixels except around the four corners of the image, and we found that the parallelization was achieved with high accuracy.

4.2 Data completion and 2D-filtering

We recorded a person walking in a room illuminated by a ceiling light in a video. He/she walked across the optical axis of the camera at a distance of about 5 [m] in going right and 4 [m] in returning with the same pace as far as possible. We applied OpenPose to get the coordinates of keypoints on the stereo-pair images in each frame as shown in **Figure 9**. Some of the joint coordinates were missing as we were afraid, so we found the sections having zero confidents and linearly interpolated them by connecting nonzero values at both ends of the sections, as shown in **Figure 10**. Top of it shows that the confidence falls to zero around 70, 90, and over 140 in frames due to

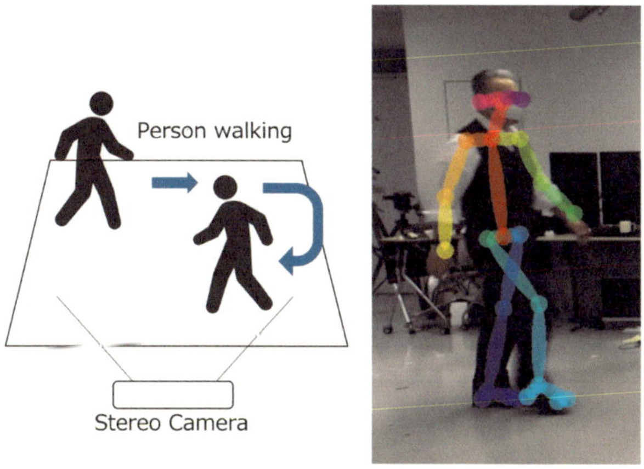

(a) *Walking path.* (b) *Keypoints recognized by OpenPose.*

Figure 9.
Walking path and an example of recognized keypoints.

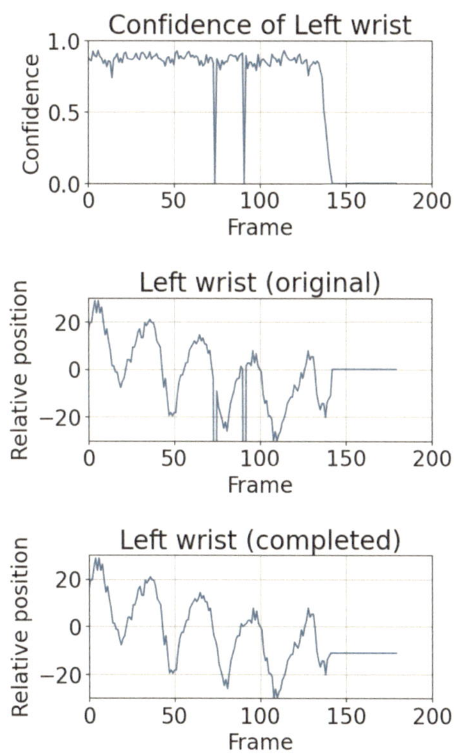

Figure 10.
Performance in completing missing data; confidence (top), missing data (middle), and completion result (bottom).

occlusions. The first two of them yield spike-shaped noises and the third zero output in the middle. The bottom shows that the spikes were well interpolated and that the last nonzero value was kept to the end of the data.

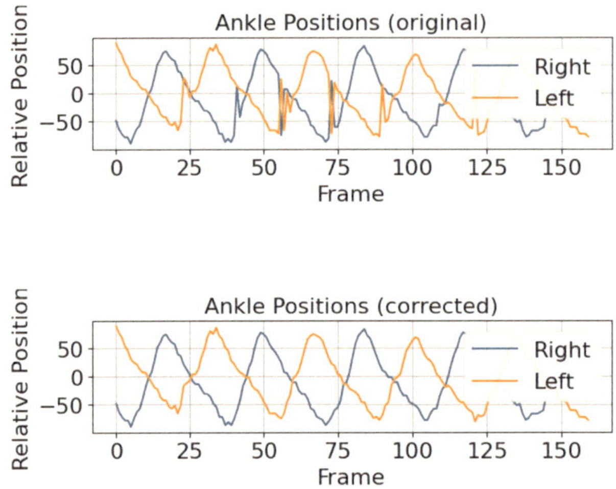

Figure 11.
Correction results for the swappings and duplications; original (upper) and correcting result (lower).

Figure 11 shows an example of the original trajectories of ankles (upper) after the missing value completion and their correction result (lower). The original trajectories included the leg duplications at frames 23–25, 41, 58, 90, and 121, and leg swappings at 56 and 73, respectively. We experimentally determined the threshold as 0.3 and got the best performance. **Figure 12** shows the performance of the low-rank filter applied to the time sequence of the right ankle horizontal coordinate. The resultant coordinates were compared with the original ones. After some trials, we set the AR model length as 12 and the rank as unity ($r_f = 1$). The low-rank filter provided us the excellent performance in reducing severe noise. The result was smooth and well reproduced the original shape, which cannot be obtained with a simple low-pass filter. Expecting to get something personal feature, we repeated the experiments by increasing the rank up to three. Unfortunately, larger ranks yielded only instability in the pose.

4.3 Puppet model fitting

Figure 13 shows the performances of the puppet model fitting for stabilizing the three-dimensional poses reconstructed from the joints on the stereo-pair images. The

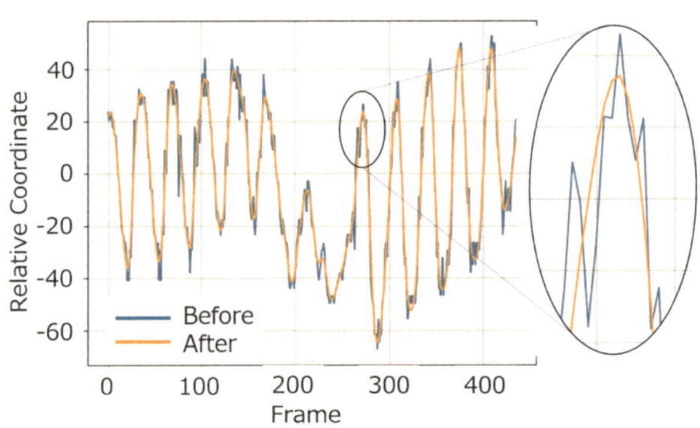

Figure 12.
Performance of 2D low-rank filter applied to the time sequence of the right ankle horizontal coordinates.

DOI: http://dx.doi.org/10.5772/intechopen.1000850

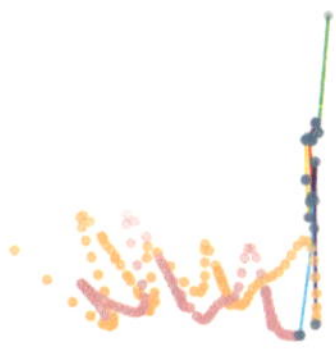

(a) *Direct reconstruction.*

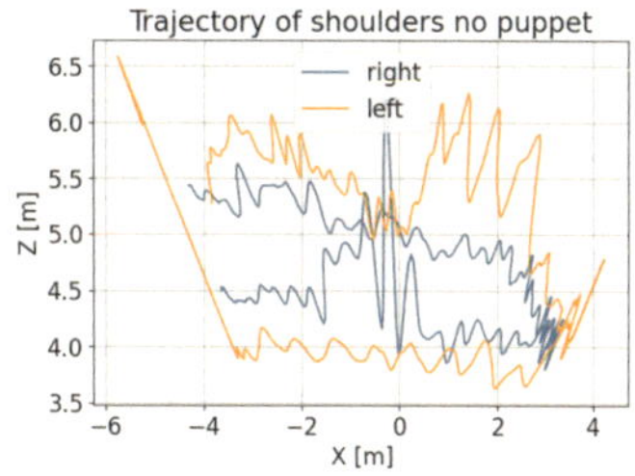

(b) *Direct trajectories.*

(c) *One with Puppet-I.*

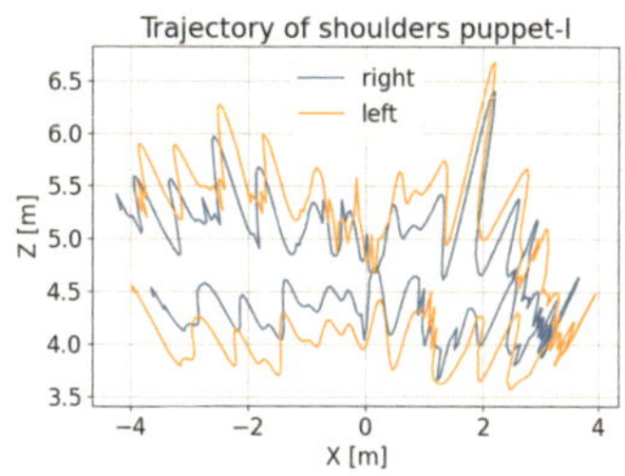

(d) *Puppet-I trajectories.*

(e) *One with Puppet-II.*

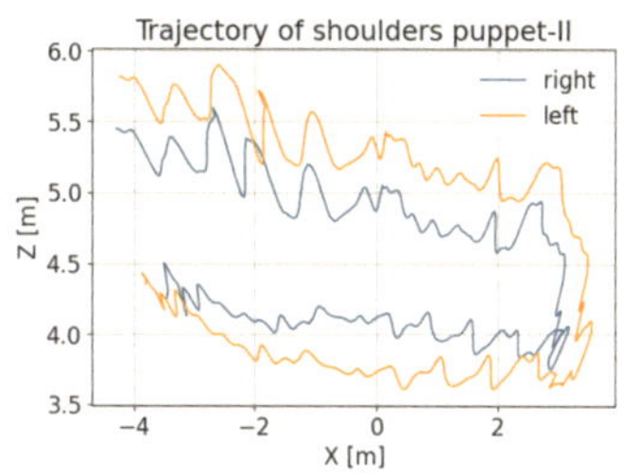

(f) *Puppet-II trajectories.*

Figure 13.
Reconstructions of 3D-pose and shoulder trajectories.

left row indicates the reconstructed pose with footprints, and the right trajectories of both shoulders are projected on the x-z plane.

In **Figure 13a**, although some joints were shifted from their original positions, their overall shape has preserved a human shape. Such discrepancies are considered to be the result of the low-rank filter in the previous stage being independently applied to each joint regardless of the mutual position of the joints. This can also be seen from the fact that the gaps between the two-dimensional trajectories of both shoulders shown in **Figure 13b** were not constant. The distortion was somewhat compensated by applying puppet-I model as shown in **Figure 13c**. The model puppet-I acted as a kind of constraint and gave the reconstructed pose a possible positional relationship among the joints as a human. You can see that the shoulder width was corrected to a constant though their positions included larger fluctuation as shown in **Figure 13d**. The fluctuation indicates that some systematic noise remains in the parallax after fitting the model puppet-I. Although the shoulder width alone is not enough, the puppet-I enabled us to reproduce relative poses from the measured joint coordinates with smoothing by the low-rank filter. We consider that the fluctuations can be reduced by applying a normal low-pass filter, such as a moving average one.

The model puppet-II was developed as the improved version of the puppet-I. This is because puppet-I rarely produces larger distortions that stretch the upper body (shoulders and arms) from the lower body (hips and legs), particularly in an area far from the camera. We speculate that this phenomenon is related to decreasing the precision in OpenPose recognition of joint positions as the object becomes smaller. Therefore, we adopted the rigid body in the puppet-II with some pose stabilizing techniques. Whereas the puppet poses are optimized in the three-dimensional world coordinate system in puppet-I, it is done on the two-dimensional image coordinate system in puppet-II. When a person crosses the optical axis and both shoulders are on it, they appear at the same point on the reference image, and the difference in their parallax is slight. Therefore, noise in the stereo-pair images prevents us to distinguish whether the front of the model is fitted to the front of the object or inverted to fit the back. Since the optimization strongly depended on the initial values, we suppressed the model flip by stabilizing them. The stabilization was performed by fitting the body model in **Figure 5a** to joint coordinates on not only the target frame but also multiple frames before and after the target frame. The multi-frame fitting was based on the fact that the corresponding point translation on the stereo-pair images with keeping the parallax just produced the translation in the world coordinate system without changing the distance. We translated the body joints on all the frames in the window so that the mid-point between the neck and the mid-hip of them matched that of the target. We set a window w with a width of 24 and used the optimization result for all frames in it as the initial values for optimization in a frame in the central small window containing 12 frames. After finishing the optimization of them, we shifted the window so that the central small window covered frames without gaps. We repeated those procedures until all frames were processed. Before performing the optimization for legs and arms, we applied an inverted test. When the direction from neck to nose in the optimized model did not match that in the measured pose, we flipped the body model. **Figure 13e** shows one of the resultant poses with footprints and **Figure 13f** shoulder trajectories. The procedures above were for adjusting the positional relationship among the joints and did not affect smoothing the distance from the camera. **Figure 14** shows the result of suppressing the variation in the distance by applying a moving average filter and their shoulder trajectories. We consider that walking human does not make sudden changes in posture for 1/3 of a second and determined the window width as 21.

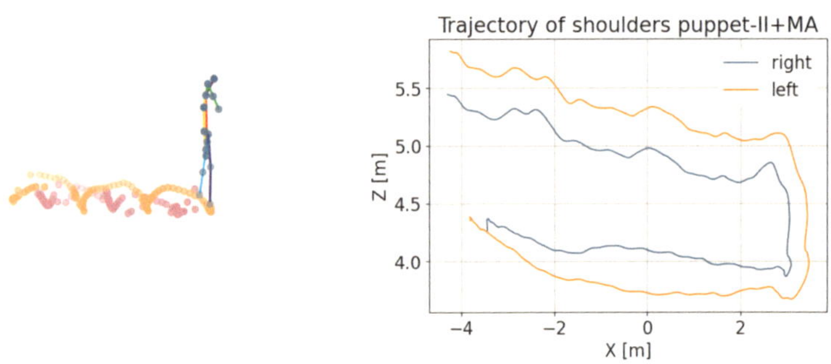

(a) *Pose with moving average.* (b) *Trajectories with moving average.*

Figure 14.
Processed 3D-pose and shoulder trajectories with moving average.

4.4 Posterial 3D-filterings

The final processings were three-dimensional low-rank filtering as a smoother in the time domain and spatial AR model filtering as one in the spatial domain. Those filters were applied to the moving average data.

Figure 15a indicates the performance of the low-rank filter LRF-3D separately applied to the time sequence of three-dimensional coordinates of the right ankle, and **Figure 15b** that of the spatial AR model filter SARF with the rank reduction from 45 to 16 using sequential 7 poses at a time applied to each pose coordinates. Both figures show the differences $dx, dy,$ and dz in coordinates $x, y,$ and z between before and after applying the filters, respectively. Pulse-shaped noises mainly in the z-coordinates were removed by applying the former filter, and those slightly remaining in the x-coordinates were cut by applying the latter one. The values on the vertical axis in **Figure 15b** were one digit smaller than those in **Figure 15a**. This indicates that the previously applied LRF-3D has removed most of the spike-shaped noise. Since the mean value of the difference did not contain a larger bias component, it is suggested that both filters worked to smooth the poses. **Figure 16** shows the final results using the models puppet-I and puppet-II, respectively. We applied the same post-processing after the puppet model fitting. In those figures, we also add the footprints and a virtual plane for recognizing the walking trajectory and for comparing their performances.

In **Figure 16a**, we see that the puppet-I produced almost the same shape in the pose itself, but slightly disturbed footprints with distorted trajectories from the U-shape. On the other hand, we see that footprints in **Figure 16b** had a more precise U-shaped trajectory and a regular pattern. These features indicate rhythmic movement of the legs in actual walking was captured well by using the model puppet-II. Therefore, we regard that the model puppet-II has higher performance as a sensing tool, especially in predicting joint positions, in the person walking measurement.

Figure 17 shows intermediate results corresponding to procedures listed in **Figure 6**. Joints reconstructed from the original data themselves were severely affected by intensive noise and gave us impossible shapes and widely scattered footprints as shown in **Figure 17b**. Applying the missing value completion did not change the situation as shown in **Figure 17c**. The impossibilities in joints' location and scattered footprints were dramatically improved by the low-rank filter LRF-2D, although some unnaturalness remained as shown in **Figure 17d**. The unnaturalness

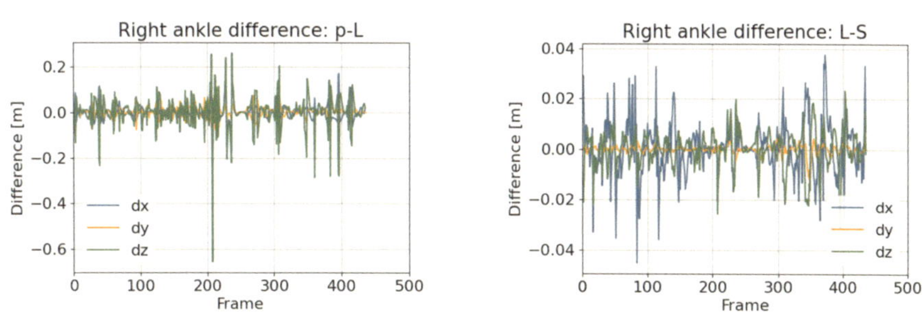

(a) *Difference between before and after the low-rank filtering.*

(b) *Difference between before and after the spatial AR model filtering.*

Figure 15.
Performance of the low-rank filter and the spatial AR model filter.

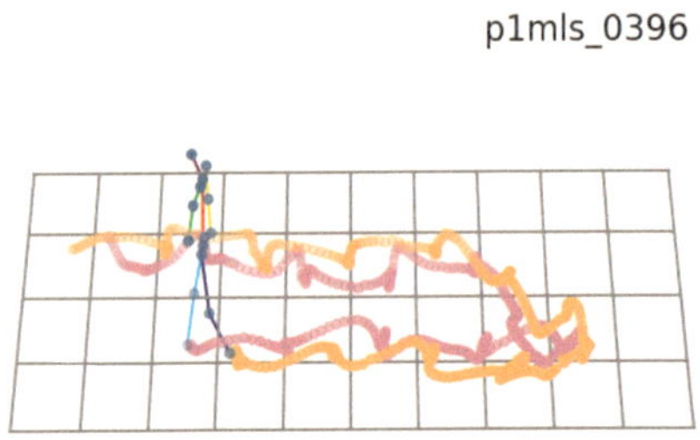

(a) *Final poses using puppet-I.*

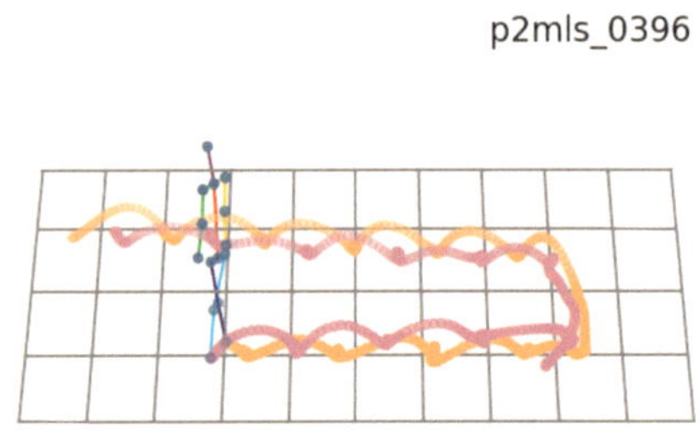

(b) *Final poses using puppet-II.*

Figure 16.
Final poses with footprints.

(a) *Observed image with joints recognized.*

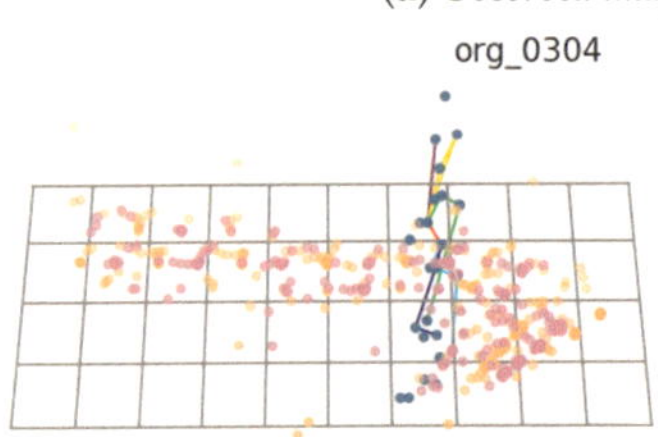

(b) *Just original.*

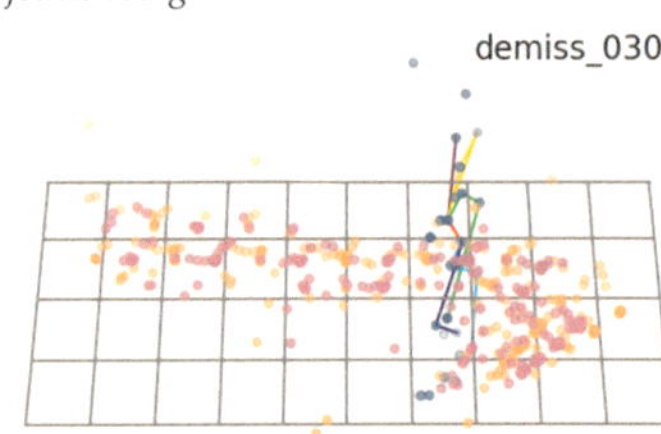

(c) *Missing value completion.*

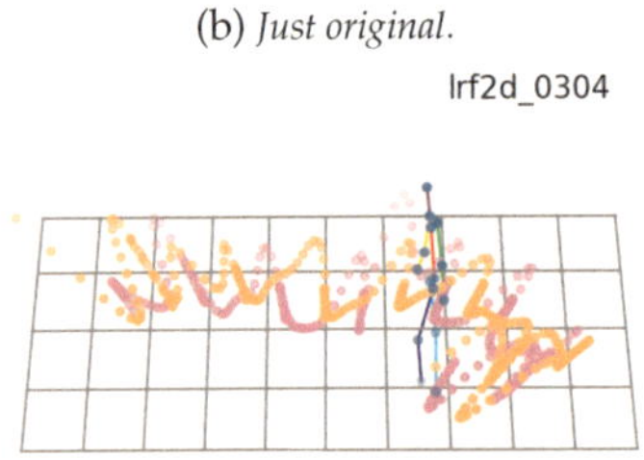

(d) *LRF-2D applied.*

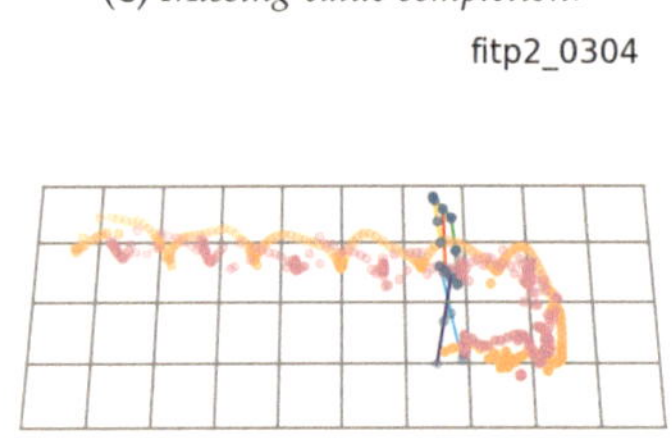

(e) *Puppet-II fitted.*

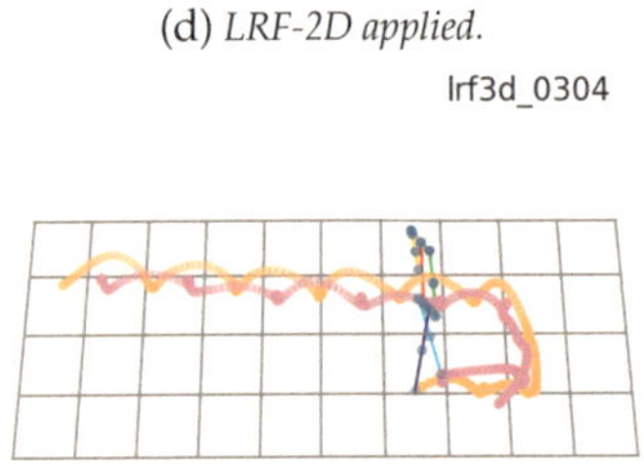

(f) *LRF-3D applied.*

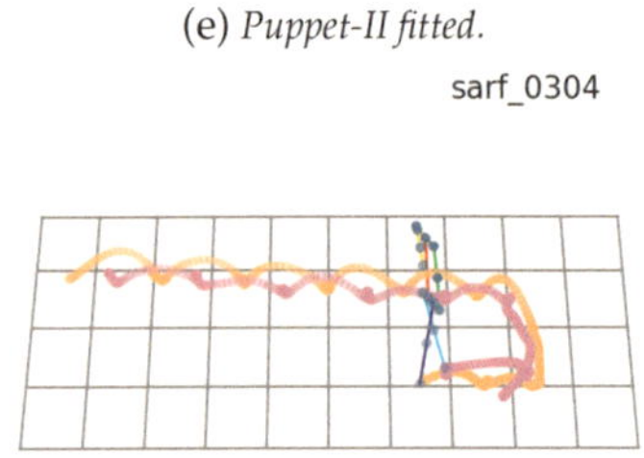

(g) *SARF applied.*

Figure 17.
Intermediate results after the main processing steps have been completed.

was removed almost perfectly by predicting and correcting body position with fitting the model puppet-II as shown in **Figure 17e**. Following the low-rank filter, LRF-3D provided smoothness in joint movements as shown in **Figure 17f**. The last spatial AR model filter SARF also gave spatial smoothness as shown in **Figure 17g**, although LRF-3D was so powerful that the effect of SARF was almost invisible. The result indicates that the proposed method measured the person walking almost perfectly in not only its pose but also its position. We also provide the whole scene of the person walking by a video at https://youtube.com/shorts/gvqF3m9xPjk. As can be seen from **Figure 16** (and the video), LRF and the puppet-II model performed very well in the walking measurements conducted in this study. The former extracts the dominant mode in the signal, so from the standpoint of the behavior measurement, the longer the same movement lasts (in this case, the more steps taken), the more stable the results will be. On the other hand, if the pose changes significantly in a short period, the dominant mode itself may become unstable and the performance may not be sufficient. We believe that such poor performance may be improved by using a stereo camera with a high frame rate for recording poses. The latter was introduced on the fact that the joint positioning in a person's upper body (shoulders and hips) does not change significantly with a pose. The good performance of this model can be interpreted as the effect of incorporating already existing knowledge into the measurement. In this case, we believe that the fixed joint positioning as a correlation structure narrows the search area in the optimizing process and thereby improves total accuracy.

Now, we consider the applicability of the model puppet-II used here to persons with different heights and body types. Based on the followings, we think the body type is not so much larger problem, and the height difference can be compensated by expanding or shrinking the length of each limb correspondingly to the height. The reasons are that roughly measured limb lengths already gave us almost the perfect results, the model had a very simple structure, and the position and the direction of body parts were determined by an optimization process, not a matching one.

Conventionally, to recognize human behavior, we have to measure poses and analyze the behavior as the temporal change of the poses. In this process, recognizing the human itself was an extremely hard task. Recent advances in deep learning have enabled software systems such as OpenPose to quantitatively extract human joints independently of the observation direction. Analysis with two-dimensional data, however, still depended on the observing direction, and one with three-dimensional data was attractive but some problems also remained in acquiring the three-dimensional human joint coordinates. In this study, we have established a technical basis for quantitatively handling three-dimensional poses and their temporal changes, that is, behavior, by removing the extreme noise in the three-dimensional joint coordinates converted from OpenPose extracting stereo-pair joint coordinates.

5. Conclusions

Based on the fact that human joints did not move at random but moved together with having a kind of spatial correlation structure in each action, a joint-based method was proposed to measure three-dimensional human pose with position in walking as the first step for obtaining the correlation structure. The joint positions were obtained by applying OpenPose to both left and right images acquired with a general-purpose stereo camera. The original three-dimensional joint coordinates were severely affected by missing values and/or positional fluctuation included in two-dimensional coordinates on the stereo-pair images. Since noise in tree-dimensional coordinates caused by Gaussian assumed fluctuation followed a Cauchy distribution, a simple averaging did not smooth the noise well. Focusing on the dominant mode, we succeeded in separating severe noise by lowering the rank of time-series data of the joint coordinates. We called it a two-dimensional low-rank filter LRF-2D. Independent application of LRF-2D to the time series of each joint coordinate yielded some positional distortions in joint locations. Those distortions were corrected by introducing the puppet model. The puppet model was introduced as a kind of restraint from the fact that lengths of the backbone, upper limbs, and lower ones were invariant in actions. Two types of model, flexible puppet and rigid one, were proposed and the rigid puppet model gave us better performance than the flexible one. In the rigid model, locations of both shoulders and hips to the backbone were fixed in the body part of the puppet, and their three-dimensional positions were determined so that the squared sum of differences among their two-dimensional joint coordinates back-projected and observed ones on stereo-pair images fell down the minimum. This meant that the optimum position of the body part was determined from observed two-dimensional joint coordinates while predicting its possible three-dimensional positions using known knowledge. Following the application of LRF-3D to time-series data of joint coordinates and application of the spatial AR model filter SARF to joints in each frame achieved almost the perfect smoothing of the joint movements in walking. The final result represented rhythmic footsteps well. Thus, the model predictive measurement derived us to the success of person walking measurement. In addition, we found that OpenPose had high precision in detecting human joints and high directional reproducibility in recognizing their positions for performing a stereo-vision measurement.

In this study, we have established a technological basis that enables a quantitative understanding of human posture and its temporal changes, that is, positioned behavior, by optical measurement from a remote location. This is expected to advance research on the recognition of human behavior and its application to gait recognition. However, although we evaluated the measurement performance of the pedestrian system qualitatively here, more quantitative evaluation is needed to use the system for applied research. At the same time, a rigorous evaluation is also needed to determine the performance of OpenPose as a human joint recognition tool, including the adoption of another recognition system with higher performance. Another future task is to extract and analyze the spatial correlation structure among all joints in movements such as walking.

Abbreviations

AR	Auto-Regression
JSON	JavaScript Object Notation
LRF	Low-rank filter
LRF-2D	Two-dimensional low-rank filter
LRF-3D	Three-dimensional low-rank filter
SARF	Spatial AR model filtering

Author details

Hiroshi Hanaizumi
Hosei University, Tokyo, Japan

*Address all correspondence to: hana@hosei.ac.jp

References

[1] Urtanstm R, Fua P. 3D Tracking for Gait Characterization and Recognitio. In: Proceedings of the Sixth IEEE International Conference on Automatic Face and Gesture Recognition (FGR'04); 17-19 May 2004; Seoul. New York: IEEE. Vol. 1. pp. 17-22

[2] Spencer N, Carter J. Towards pose invariant gait reconstruction. In: Proceedings of IEEE International Conference on Image Processing 2005; New York: IEEE. Vol. 3. 2005

[3] Bashir K, Xiang T, Gong S, Mary Q. Gait representation using flow fields. In: Proceedings Of the British Machine Vision Conference 2009 (BMVC 2009); London. 7-10 September 2009

[4] Munea TL, Jembre YZ, Weldegebriel HT, Chen L, Huang C, Yang C. The progress of human pose estimation: A survey and taxonomy of models applied in 2D human pose estimation. IEEE Access. 2020;**8**: 133330-133348

[5] Khamsemanan N, Nattee C, Jianwattanapaisarn N. Human identification from freestyle walks using posture-based gait feature. IEEE Transaction on Information Forensics and Security. 2018;**13**(1):119-128

[6] Gianaria E, Balossino N, Velonaki M. Gait characterization using dynamic skeleton acquisition. In: Proceedings of the IEEE 15th International Workshop on Multimedia Signal Processing (MMSP 2013); September 30–October 2 2013; Pula. New York: IEEE. 2013. pp. 440-445

[7] Chalottopadhyay P, Sural S, Mukherjee J. Frontal gait recognition from incomplete sequences using RGB-D camera. IEEE Transactions on Information Forensics and Security. 2014;**9**(11):1843-1856

[8] Ahmed F, Paul PP, Gavrilova ML. DTW-based kernel and rank-level fusion for 3D gait recognition using Kinect. The Visual Computer. 2015;**31**:915-924. DOI: 10.1007/s00371-015-1092-0

[9] Cao Z, Hidalgo G, Simon T, Wei S, Sheikh Y. OpenPose: Realtime multi-person 2D pose estimation using part affinity fields. IEEE Transactions on Pattern Analysis and Machine Intelligence. 2021;**43**:172-186

[10] Hanaizumi H, Misono H. An openpose based medhod to detect texting while walking. In: Proceedings of The 7th International Conference on, Intelligent Systems and Image Processing 2019 (ICISIP 2019); Taipei; 5-9 September 2019. pp. 130-134

[11] Konishi K, Uruma K, Takahashi T, Furukawa T. Iterative partial matrix shrinkage algorithm for matrix rank minimization. Signal Processing. 2014; **100**:124-131. DOI: 10.1016/j.sigpro.2014.01.014

[12] Hanaizumi H, Otahara A. A Method for Measuring Three-Dimensional Human Joint Movements in Walking. In: Proceedings of the 60th Annual Conference of the Society of Instrument and Control Engineers of Japan 2021 (SICE 2021); Tokyo; 8-12 September 2021. pp. 1476-1481

[13] Wikipedia: Ratio distrubution [Internet]. Available from: https://en.wikipedia.org/wiki/Ratio_distribution [Accessed: 2022-11-30]

Chapter 4

Perspective Chapter: Model Predictive Control of Multi-Variable Non-Linear System Using Multi-Agent Q-Learning

Song Ho Kim and Kwang Rim Song

Abstract

Model predictive control (MPC) is generally implemented with the feedback controller to calculate the control sequence by solving an open-loop optimization problem in the finite horizon, subject to the input and output constraints in the direct control layer. While predicting the dynamics of the non-convex and non-linear system, the computational burden and the model uncertainty are primary difficulties of the MPC implementation. The objective of this paper is to design a hierarchically structured MPC as the online set-point optimizer for maximizing economic efficiency over a long time horizon to meet the strict product quality requirements. This paper proposes a new method of using cooperative actions of multi-agent Q-learning for determining the set-point of controllers to approximate the temperature trajectory of the raw material through the cascaded three-step kiln process to an integrated reference trajectory. Experimental results show that these cooperative actions guarantee the online set-point optimization for decentralized local controllers within the framework of hierarchically structured MPC.

Keywords: model predictive control, cooperative actions, online set-point optimization, multi-agent, Q-learning

1. Introduction

Model predictive control (MPC) is a control method that the future response of a plant is predicted by the process model and the set of future control signals is calculated by optimizing a determined criterion to keep the process as close as possible to the reference trajectory. MPC has originated from engineering requirements to maximize the control performance and the economic efficiency of the complex processes, where requirements for manufacturing could not be properly satisfied by only the feedback control such as PID.

MPC has been successfully implemented with a specific named controller for almost all kinds of systems such as industry, economics, management and finance [1].

Nowadays, MPC is also referred as model-based predictive control, receding horizon control (RHC), or moving horizon optimal control. MPC has become one of the e advanced process control technologies widely used for practical applications such as multivariable control problems in the process industry because of its own characteristics to consider the physical and operational constraints.

In traditional model predictive control, the control action at every sampling time is determined by solving an online optimization problem with the initial state, being the current state of the plant. MPC adopts the open-loop control method, by which the future control actions are computed to minimize the performance function subject to the desired reference trajectory over a prediction horizon under the constraints on input and output signals at every sampling time. MPC has some features such as compensating large delay, considering physical constraints, handling multivariable system and not requiring deep knowledge of control. Especially, one of the key features of MPC is the ability to handle constraints in the design [2, 3].

In most MPC algorithms, the predictive model would be the state space model. Since the state space model deals with the state directly, MPC could be expanded to multivariable systems and implement optimal control considering state constraints. However, MPC also has some drawbacks such as the high computational burden and the stringent real-time requirement, because it has to solve iterative optimal problems at every sampling time.

In many research literatures for industrial processes control, MPC was designed as the feedback controller to calculate the control sequence by solving a finite horizon open-loop optimal control problem, subject to constraints on input and output signals at direct control level. However, in the light of implementation, the computational burden to calculate optimization problem at every sampling time remains the most critical challenge. Moreover, the model uncertainty to predict the non-convex and non-linear dynamic systems lead also to difficulties.

To cope with the non-linearity of the industrial processes, several non-linear MPC techniques have been developed for multivariable optimal control using a non-linear dynamic model for the prediction of the process outputs [4]. Two methods are introduced which are based on non-linear model predictive control (NMPC) to solve stochastic optimization problems, considering uncertainty and disturbances [5].

The distributed model predictive control is proposed, which is the effective approach for plants with several subsystems [6, 7]. However, the direct control might be insufficient for the optimal controller design in complex multivariable processes due to their strong uncertainty, non-linearity and time-varying disturbance inputs. To deal with these complex situations, a hierarchical control structure is presented to maximize economic efficiency in the long run of the process so that it is possible to efficiently handle the complexity and uncertainty in the process dynamics [3, 8].

MPC has been called the set-point optimizer in the constrained control layer (or the global control layer) of the hierarchical structure. And it determines an optimal set-point which leads to the best-predicted output of the process over some limited horizon for the feedback controller of the direct control layer (or local control layer).

The multilayer control structuring methodology with the control layer in the complex industrial processes was proposed in [8], and the online set-point optimizer has been designed using a hierarchical control structure with MPC advanced control layer in [9, 10]. The hierarchically structured MPC determining set-points for local controllers placed at different elements of the sewer network have been designed in [2] and the hierarchically distributed MPC that a higher supervisory layer provides set-point trajectories to lower level MPCs proposed in [11].

As MPC in feedback controller, the optimizer of set-point control layer in the hierarchical structure needs also the explicit model on processes. Nevertheless, it is generally very difficult to obtain the dynamic model of the complex controlled plant. Therefore, Reference [12] proposed a Q-learning-based intelligent supervisory control system (ISCS) with two layer-structures to find the online optimal set-points of control loops for the three-step kiln process and presented a method for improvement of Q-learning convergence to solve the difficulties of obtaining precise models of industrial processes. By extending this method, this paper proposes a new method that exploits the cooperative action of the multi-agent Q-learning, which satisfies the online set-point optimization of the decentralized controllers to approximate the temperature trajectory of the raw material of passing through the cascaded three-step kiln to an integrated reference trajectory.

The objective of this paper is to design a hierarchically structured MPC as online set-point optimizer for maximizing the economic efficiency over a long time horizon to meet the strict quality requirements for products. The rest of this chapter is organized as follows. Section 2 designs the distributed MPC system of a three-layer structure, summarizes the function of every layer and analyze the requirement for the controlled plant and constraints. Section 3 designs the Q-learning agents for the online set-point optimization of the decentralized controllers and Section 4 describes the methodology to improve the Q-learning convergence for the intelligent supervisory control of the kiln process. Section 5 describes the simulation and practical results to demonstrate the effectiveness of the proposed method. Finally, we draw conclusions and give some directions for future research.

2. Design of MPC system with hierarchical structure

This section describes the design and functions of the hierarchically distributed MPC architecture for set-point optimization.

2.1 System architecture

The kiln process consists of three electric heating zones, namely the preheating zone, the heating zone and the reducing zone, which are cascaded. In this process, the quality requirements for most of the products were met by only stabilizing the color of the product, which can be determined by the overall quantity of heat permeating through materials. The temperature and speed in the heating zone allow us to control the heat quantity, provided that the temperature of the preheating and reducing zone was kept constant. Therefore, the intelligent supervisory control system of a two-layer structure based on Q-learning has been proposed, which determines the set-point of the controller in the heating zone instead of the human operator [12].

However, some specific products require not only the color property but also stricter physical properties which depend on the temperature change of the preheating and reducing zones. To do this, the temperature change characteristics of the material passing through each zone should be approximated to a given reference trajectory.

On the other hand, the kiln process is a multivariable non-linear system in which the quality characteristics of the system cannot be evaluated by the response to any single input signal alone, and vary with the disturbance. Moreover, it may be very difficult to satisfy the quality characteristics of the high control requirements only by linear feedback control, where the three zones in the process control are approximately linearized, respectively. Therefore, this paper extends ISCS of the two-layer structure in the framework of MPC to ensure the temperature trajectory of the raw material passing through all zones in the kiln and proposes a three-layer ISCS architecture with multi-agent for the set-point optimization of the decentralized local controllers.

2.2 Kiln process and control requirements

Since the sizer followed by kilns classifies the material to flow into three lines, the quantity of input material differs from each other. The experiment shows that the average variation cycle of the material quantity is 1.2 min and the variation rate is 13.4%.

Passing through the preheating zone, the raw materials are slowly heated along with a moderate temperature rise trajectory, depending on the requirements of the physical initial properties of the product. Then, the temperature change of the materials depends on the time of passing through this zone and its temperature, which are determined by the set-point of the control system. When the preheating zone temperature is stabilized according to the set-point by PID control, the heat transfer characteristics of the raw material passing through this zone seem to be approximated by the first-order element with time delay from the step response experiment. **Figure 1(a)** shows the step response curve and the transfer function is equal to Eq. (1).

$$G(s) = \frac{k}{\mathrm{Ts}+1}e^{-\tau s} \tag{1}$$

where T is the heat transfer time constant depending on the mass of the material, τ is the time delay, and k is the steady-state value.

As shown in **Figure 1**, the value of the steady-state k impacts on the reachable temperature at a certain period. The elapsed time of passing through the preheating zone also varies depending on the conveyor moving speed. Therefore, in order to satisfy the temperature reference trajectory of the raw materials, it is necessary to change the temperature set-point of this zone according to the conveyor speed. The heating zone is the most important one for ensuring the physicochemical properties of products. The color of products that reflects the primary quality requirements is changed sensitively by the temperature, conveyor speed and input material quantity in this zone. Therefore, to achieve the desired color of products, it is necessary to change the temperature and conveyor speed of this zone with the variation of material

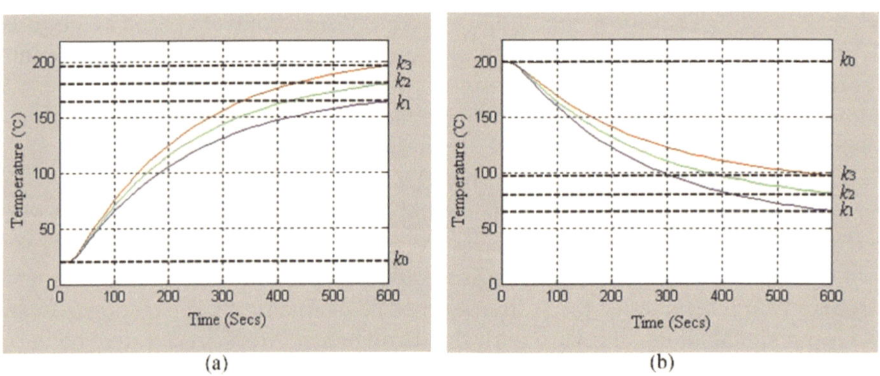

Figure 1.
Step responses. (a) Preheating zone. (b) Reducing zone.

quantity. The conveyor speed of the kiln process is controlled to only satisfy the control characteristics of the heating zone under certain constraints, and the speed in the preheating and reducing zone is dependent on the heating zone.

In the reducing zone, the materials are cooled slowly along with the moderate temperature drop trajectory from the outlet temperature in the heating zone to meet the physical properties according to the product quality stabilization requirements. The temperature change of the materials depends on the temperature and passing time of this zone.

The temperature change of the materials and control requirement in this zone is the same as the preheating zone, but the only difference is that this is the cooling process due to the temperature drop. The heat transfer model is the same as for the preheating process and the step response is shown in **Figure 1(b)**.

Each zone and conveyor of the kiln are controlled independently by a decentralized PID controller according to a given set-point, respectively. In **Figure 2**, the set-points are denoted as $w_1(s)$, $w_2(s)$, $w_3(s)$, and $w_4(s)$; manipulated variables as $u_1(s)$, $u_2(s)$ and $u_3(s)$; the controlled variables as $y_1(s)$, $y_2(s)$ and $y_3(s)$ and v, which respectively correspond to the temperature T_1, T_2 and T_3 and the conveyor speed. If the quantity of the material in the flow lines remains constant, the thermal equilibrium would be also maintained. If not, it is possible that the thermal balance in the kiln would be affected by the change in the material quantity. Therefore, the overall heat quantity of the materials is calculated as follows:

$$Qc = Qc_1 + Qc_2 + Qc_3 \text{ (kJ)} \tag{2}$$

where Qc_1, Qc_2 and Qc_3 are the heat quantity of each zone.

Since the variation of the heat quantity due to the change of the material quantity causes the variation of the surface color of the products from kilns, it is unlikely that the color characteristics of the products would be stabilized by the control loops.

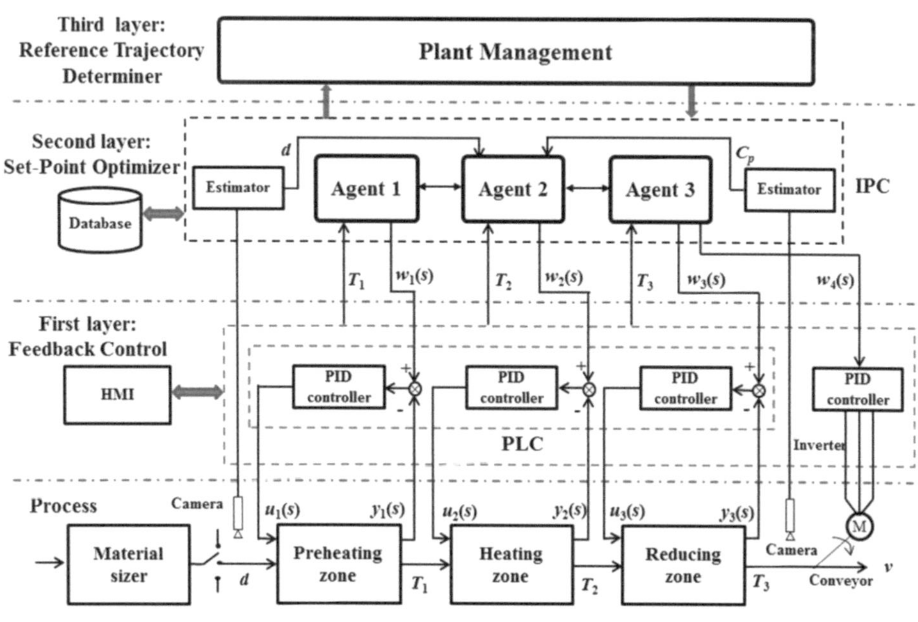

Figure 2.
Three-layer ISCS architecture.

2.3 MPC: Set-point optimization

The two-layer structured ISCS, which employs the reinforcement learning method in which the agent learns through interaction with the environment, was developed to optimize the set-points of the temperature and speed controllers instead of the human operator in the heating zone [12].

The human operator empirically determines the set-points of the control loops in the original process and evaluates the input material quantity and the color of product images by using CCD cameras at the input and output of kiln processes. After that where set-points for the temperature and the conveyor belt speed in each zone are properly provided through HMI (Human Machine Interface) for maintaining the product color to the required value.

Due to the difficulty of considering all the variables in the operation of the real process, the human operator usually regulates set-points of the conveyor speed and the temperature for the heating zone to cope with the change of the material quantity, but not the temperature for the preheating and reduction zone. During the process operation, all values of process variables are stored in the database.

Moreover, considering the fuzziness of the human's action for product stabilization, the following relationships between state and action variables would hold:

$$\text{IF } d = \text{A and } v = \text{B and } T_i = \text{C}_i \text{ THEN} \Delta v = \text{E and} \Delta T_i = \text{F}_i \tag{3}$$

where d (kg/m^2) is the quantity of material, v (m/s) is the speed of the conveyor, T_i (°C) is the temperature of the heating zone and Δv and ΔT_i are the offset of the speed and the temperature set-points, and A, B, C_i, E and F_i are fuzzy sets.

The proposed three-layer structured ISCS consists of a distributed MPC system with several agents, in which the set-point optimizers of the second layer calculate the optimal set-points of the decentralized controllers to satisfy the control characteristics according to the product quality requirements in each zone with the reference trajectory provided by the economic optimization layer.

Agent 1 receives the temperature T_1 of the preheating zone and the conveyor speed v determined by agent 2 as stated and determines the temperature set-point change ΔT_1 of a controller in accordance with the quality requirements in this zone as an action. Thus, the action of agent 1 makes a decision to satisfy the economic efficiency by depending on the action of agent 2.

Agent 2 observes states such as speed v, temperature T_2 in the heating zone and the input material quantity d obtained by the estimator from CCD camera images, and then, it determines the optimal change of set-point Δv and ΔT_2 for speed and temperature of the controller to stabilize the output product color C_p. Agent 2 does not depend on the action of other agents, acts independently, and only provides its own action results upon their request.

Agent 3 receives the speed v and the temperature T_3 of the reducing zone as stated and determines the temperature set-point change ΔT_3 of the controller as an action to meet the quality requirements. The action of agent 3 also depends on the action of agent 2 as well as agent 1.

This system can be modeled as a multi-agent system, which is a group of single agents with independent subsystems. Reinforcement Learning is a powerful approach to solving multi-agent learning problems in complex environments.

This paper focuses on the design problem of multi-agent Q-learning for set-point optimization of controllers in the set-point optimization layer to solve the distributed

NMPC problem, which approximates the temperature trajectory of materials to the reference trajectory in each zone.

The distributed NMPC problem is to determine the temperature set-point change ΔT_i and the speed set-point change Δv, which minimizes the following cost functions $J_i(k)$ subject to constraints on input and output signals in each zone at the sampling time k:

$$J_1(k) = \sum_{p=1}^{L-1} (r_1(k+p) - y_1(k+p))^2 + \sum_{p=0}^{L-2} \Delta T_1(k+p)^2 \tag{4}$$

$$J_2(k) = \sum_{p=L}^{M-1} (r_2(k+p) - y_2(k+p))^2 + \sum_{p=L-1}^{M-2} \Delta T_2(k+p)^2 + \sum_{p=L-1}^{M-2} \Delta v(k+p)^2 \tag{5}$$

$$J_3(k) = \sum_{p=M}^{N} (r_3(k+p) - y_3(k+p))^2 + \sum_{p=M-1}^{N-1} \Delta T_3(k+p)^2 \tag{6}$$

where N is the total predicted length (p = 1, 2, ..., N), y_i is the temperature prediction trajectory of materials and r_i is the reference trajectory. That is, NMPC is formulated as the optimal control problem with the objective function as follows:

$$\min J(k) = \sum_{i=1}^{3} J_i(k) \tag{7}$$

subject to:

$$\begin{gathered} T_{i,\min} \leq T_i(k) \leq T_{i,\max} \\ \Delta T_{i,\min} \leq \Delta T_i(k) \leq \Delta T_{i,\max} \\ v_{\min} v(k) v_{\max} \\ \Delta v_{\min} \Delta v(k) \Delta v_{\max} \\ y_{i,\min} \leq y^i(k) \leq y_{i,\max} \end{gathered} \tag{8}$$

With constraints, the cost functions can be minimized numerically by the optimal action of agents.

2.4 Operational objectives determination

The third layer functions as the plant control layer, where the optimal reference trajectory would be calculated to achieve the economic effectiveness of the process

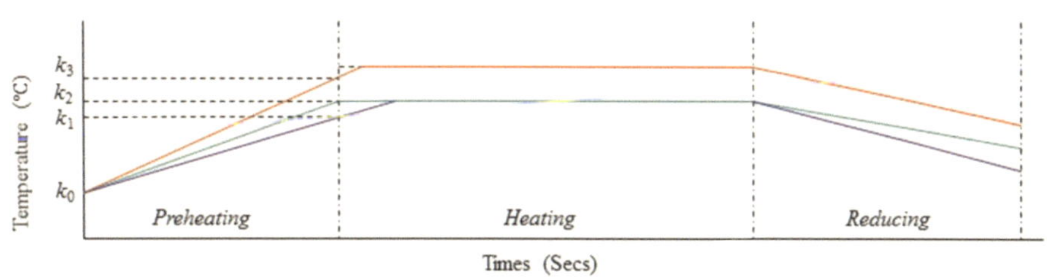

Figure 3.
The reference trajectories for three-step kiln process.

according to its operational objective determined by the quality requirement of the product. The obtained reference trajectory would be provided to the decentralized agents. **Figure 3** shows three typical reference trajectories determined according to the operating objectives of the quality requirements in each zone.

3. Design of Q-learning agents for online set-point optimization

This section describes the design of the decentralized Q-learning agents for set-point optimization of controllers in each zone to solve the distributed NMPC problem.

3.1 Q-learning and multi-agent

From the analysis of references [13, 14], Q-learning is described as follows [12].

The Q-learning algorithm is a model-free reinforcement learning method and its convergence has been proved. The Q-learning estimates the expected value $Q^*(s, a)$ of the discounted reward sum as Q-value through randomized interactions with the environment without the requirement of knowledge for transition probabilities in Markov decision process (MDP).

The action value $Q(s, a)$ is the estimate of an action a in a state s and its updating process would be recorded in a Q-table. Generally, the initial value of the $Q(s, a)$ called Q-function is given randomly and is approximated iteratively to the optimal Q-function. As a state transits from s_t to s_{t+1} by an action a, the agent updates $Q(s_t, a_t)$ as follows:

$$Q(s_t, a_t) \Leftarrow (1-\alpha)Q(s_t, a_t) + \alpha\left[r_{t+1} + \gamma \max_a Q(s_{t+1}, a)\right] \quad (9)$$

where $\alpha \in (0, 1]$ is the learning rate, r_{t+1} the immediate reward for the next state s_{t+1} and $\max_a Q(s_{t+1}, a)$ the maximum action value in the same state. The Q-learning algorithm for estimating $Q^*(s, a)$ is given in **Table 1**.

The Q-table can be utilized as a dynamic model of the environment after the learning has been completed because it reflects the dynamic characteristic of the environment.

Despite the effectiveness for optimal policy, the Q-learning algorithm has some drawbacks of slow convergence. To tackle these issues, problem definition, discretization and learning parameters should be properly determined in practical circumstances [15].

It is usually difficult for a single agent to implement a large-scale complex task due to its own limited capacity. Multi-agents, which collaborate agents with each other

1. Set the initial value of $Q(s, a)$ randomly and $t = 0$.
2. Measure the current state s_t.
3. Determine an action a_t from s_t using policy given in Q-table.
4. Apply action a, measure next state s_{t+1} and reward r_{t+1}.
5. Update $Q(s, a)$ by Eq. (9).
6. $t \leftarrow t + 1$.
7. Repeat from step 2 until convergence.

Table 1.
Q-learning algorithm.

through decentralized computation, can greatly improve their solving efficiency. However, since multi-agent learning process is quite complex, it is not so simple to solve cooperative problem of single agents.

The cooperative control problem of decentralized multi-agents for complex objects that cannot be solved by a single agent has been studied [16–18].

Although multi-agent systems are effective in large-scale complex environments, many difficulties remain in their solution methods.

3.2 Agent design for heating zone

The heating zone is the most important part for the quality of products, and the agent for the determination of the controller set-point in this zone does not depend on the action of the other agents, but only transmits its future action through the message with the request of other agents. Thus, the agent design in this zone becomes the single agent problem. The materials passing through this zone cause physicochemical changes and are calcined near the appropriate temperature set-point for a certain time.

Although the temperature in this zone is at a steady state, the temperature trajectory of materials cannot be properly predicted by the heat transfer model because the input material quantity changes the variation of the heat quantity of materials. Hence, in order to stabilize the heat quantity that materials receive, the agent evaluates the color of the output product and determines the optimal set-point change Δv and ΔT_2 of the temperature and conveyor speed for the heating zone controller according to Eq. (3) in each action cycle. Thus, Eq. (5) is reformulated as follows:

$$J_2(k) = \sum_{p=L}^{M-1} \left(r_{\text{cp}}(k+p) - y_{\text{cp}}(k+p) \right)^2 + \sum_{p=L-1}^{M-2} \Delta T_2(k+p)^2 + \sum_{p=L-1}^{M-2} \Delta v(k+p)^2 \quad (10)$$

where r_{cp} is the desired reference luminance of the product, and y_{cp} is the measured luminance obtained in the estimator.

The agent designing for the heating zone involves the designing of the state space, action space and reward function as in [12] (see Appendix A).

3.3 Cooperative agent design for preheating and reducing zones

Like single-agent reinforcement learning, multi-agent reinforcement learning is also based on MDP as a formalization of multi-agent interaction. Agents act collectively and cooperate to achieve their own individual goals as well as the common goal of the group to which they belong. However, each agent often cooperates with the other to achieve their own individual goals rather than solving a common problem.

In fact, since all the agents make their decisions simultaneously, each agent cannot recognize the others' choices in advance, but make a choice to possibly transit from its current state to a better state. Therefore, the paper proposes multi-agent Q-learning as a form of joint state and individual action.

Preheating and reducing are the processes to satisfy the physical properties according to the specific quality requirements of the product. In each zones, agents 1 and 3 predict the temperature change y_i of the material according to the current temperature T_i (i = 1, 3) and the conveyor speed v in each action cycle using the step response model over the prediction horizon and calculate the optimal set-point changes of the controllers from Eq. (4) and (6) to approximate them to the reference trajectory r_i, respectively.

Since the conveyor velocity v is determined by the heating zone agent 2, the action of agents 1 and 3 depends on the action of agent 2. That is, the Q-learning agents act to maximize the reward based on the cooperative information with agent 2 and the independent interaction with the sub-environment. The agent's action is represented by the following fuzzy rules:

$$\text{IF } v = \text{B and } T_i = \text{C}_i \text{ THEN} \Delta T_i = \text{F}_i, i = 1, 3 \tag{11}$$

where B, C_i and F_i are fuzzy sets of physical quantities corresponding to linguistic evaluation.

3.3.1 State space

From Eq. (11), states of the preheating and reducing zones are the joint state of the state v of agent 2 and a state T_i of agent i (i = 1, 3). The continuous state variables v and T_i are discretized and the state space is partitioned into fuzzy subspaces that are assigned to five linguistic variables, respectively. The discrete joint state space can be represented as follows:

$$S = \{s|s_{jk} = (v_j, T_{i,k}), i = 1, 3, j = 1, 2, \ldots, J, k = 1, 2, \ldots, K\} \tag{12}$$

where $J = K = 5$ and the number of possible states are $5^2 = 25$ and the triangular membership functions are used.

3.3.2 Action space

As shown in Eq. (11), the set-point change ΔT_i of controllers is the action variable of the agents for the operation of the preheating and reducing zones. The range of this variable is determined by $\Delta T_i \in [-15, +15]$ (°C) as the heating zone. The continuous action space is partitioned into fuzzy subspace which is assigned to five linguistic variables. The discrete action space for the agent is represented as follows:

$$A = \{a|a_m = \Delta T_{i,m}, i = 1, 3, m = 1, 2, \ldots, M\} \tag{13}$$

where M is the number of equally divided fuzzy subset in the proper range of $\Delta T_{i,m}$.

In each time step, based on the observed states (i.e. v_j and $T_{i,k}$) and the ε-greedy policy, the agents choose a state-action pair from Q-table and update the temperature set-points of control systems for speed and temperature for the preheating and reducing zones. By the action of agents, the temperatures are updated as follows:

$$T_i(t+1) = T_i(t) + \Delta T_i(t+1), i = 1, 3 \tag{14}$$

3.3.3 Reward function

The action of agents in the preheating and reducing zones to meet the specific quality requirements of the product is guaranteed by approximating the heat transfer trajectory of the material passing through these zones to the reference trajectory.

Thus, the reward signal of Q-learning for the control of these zones can be expressed as the inverse of the square sum of the deviation between the reference trajectory and the prediction trajectory over the prediction horizon, which is obtained by the set-point change at the current sampling time as follows:

$$r_{t+1} = \frac{1}{\sum_{k=0}^{K} \left(r_i(k) - y_i(k)\right)^2}, i = 1, 3 \tag{15}$$

where k is the prediction length in the ith zone and $\sum_{k=0}^{K} \left(r_i(k) - y_i(k)\right)^2 > 0$.

4. Q-learning convergence improvement

This section describes the convergence improvement of Q-learning agents designed above. Firstly, the results of our study on how to improve the convergence of Q-learning in the heating zone are presented. Next, to improve the convergence of Q-learning in the preheating and reducing zones, the initialization method of Q-table using the model is described.

4.1 Convergence improvement method of Q-learning using operational data

Methods for improving the convergence of Q-learning with discrete action as the primary acting mode are largely divided into tuning the learning parameters and fast updating of the Q-table.

In the heating zone, the convergence improvement method [12] of the Q-learning agent based on the method of obtaining the optimal Q-table is presented in Appendix B, but a brief outline is as follows.

The use of the agent in process operation has not been permitted until the learning is completed, because the heating zone agent uses a model-free Q-learning method that finds the optimal policy without prior knowledge of the environment.

Therefore, the method for improving the convergence of Q-learning is to determine the initial values of the Q-table optimally. This requires knowledge of the environmental dynamics, which is not easily obtained in the non-linear system.

The proposed method is realized by extracting automatically the operational experience rules of the expert from past operating data and initializing the Q-table using it.

4.2 Q-table initialization using model

In the zones for preheating and reducing of the calcined raw material, agents interact with the environment and adjust the set-point of the control systems according to the conveyor speed change, considering the cooperative information provided by the heating zone agent.

Taking into account the arbitrary variation of the speed, the agent action of determining the set-point change with the temperature of its zone is represented by a fuzzy logic rule as Eq. (11). When two state variables and one action variable have five fuzzy variables, respectively, the Q-table consists of 125 state-action pairs where the number of states is 25 and each state has 5 actions, respectively.

Rule	Description
R1	x_1 = NB & x_2 = NB & y = PB ⇨ Q = 9.8
R2	x_1 = NB & x_2 = NM & y = PM ⇨ Q = 7.6
R3	x_1 = ZO & x_2 = ZO & y = ZO ⇨ Q = 11.2
…	…
R125	x_1 = PB & x_2 = PB & y = NB ⇨ Q = 1.4

Table 2.
Structure of rules.

In the preheating and reducing zones, the heat transfer dynamics of the material from the step response experiment are shown in **Figure 1**. Given the dynamics, it is possible to perform the offline simulation.

This subsection presents a method of determining the optimal initial value of the Q-table using the dynamics model. Given the conveyor speed v determined by the action of the heating zone agent, the time t of passing through the zone can be calculated.

Based on the numerical simulation using the step response model, moreover, the set-point change where the heat transfer trajectory of the material is approximated to the reference trajectory can be determined repeatedly.

First, the initial Q-value is set to 0 and the reward is determined by the off-line simulation using the step response model and Eq. (15).

Eq. (15) is the reward function at a joint state S and an individual action A of agent i, that is,

$$R_i : S \times A \rightarrow \mathbf{R} \tag{16}$$

where **R** is the set of real numbers. **Table 2** shows the structure of rules.

5. Results and discussion

The simulation result to evaluate the performance of the agent for the heating zone is presented in Appendix C.

The simulation result for the preheating and reducing zones does not present because there have already existed the explicit heat transfer models in these zones, but only the operating experiment result.

During the operation experiment, firstly, the luminance of the output product is utilized to compare the operation by the human with the agent for the heating zone [12].

It is usually impossible to run the comparative experimentation with the traditional method in a real situation due to the trial-and-error features in the Q-learning with arbitrarily initialized Q-value. Therefore, the operations of an operator are preferred to evaluate the proposed method.

Figure 4 shows the comparison results for 90 hour-experiment.

The tolerance limitation Y_0 of luminance lies between 167 and 178. The luminance Y for the manual operation changes within the range of $Y_0 \pm 5$ and its deviation V is 0.4%. For the proposed case, the experiment results are that Y=$Y_0 \pm 3$ and V = 0.4%.

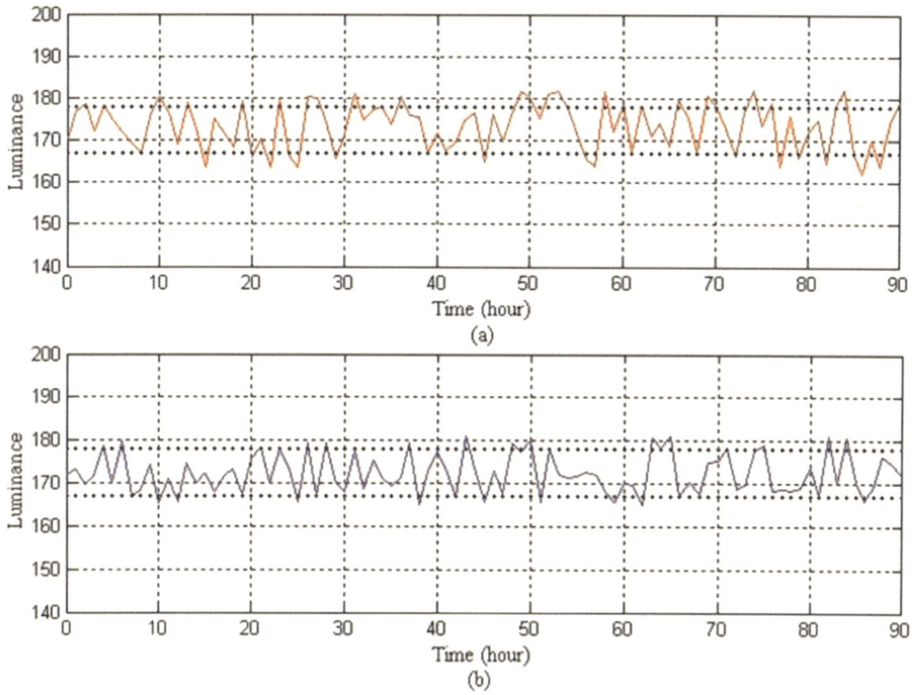

Figure 4.
Operation experiment result of real process. (a) Operation by human operator. (b) Operation by agent.

In spite of the sub-optimality of the initial Q-values, it seems that the convergence feature was mostly satisfied.

However, the experiment for human operation shows somewhat insufficiency. The operator's experience might have a decisive impact on the long-term operation. Furthermore, it is unlikely that human might always make the correct decision for set-points of control systems.

From the above results, it is certain that the proposed method improves the trial-and-error features in the Q-learning with arbitrarily initialized Q-value.

Next, in the preheating and reducing zones, the online set-point determination method using the cooperative action of agents is compared with the previous method by the deviation between the predicted temperature trajectory and the reference trajectory.

Figure 5 shows the results of an experiment for the comparison of the temperature change trajectory with the reference trajectory according to the operating mode of the kiln process. Since the temperature trajectory of the material in the heating zone cannot be predicted by the heat transfer model, the medium part of the plot for this zone demonstrates the relative change of the heat quantity estimated by the luminance of the product during the experiment.

In the previous approach, the supervisory control by the intelligent agent was implemented for the heating zone with the temperature of the preheating and reducing zones being constant due to the operational complexity. In this operational mode, the reference trajectories and the predicted temperature trajectories are shown in **Figure 5(a)**.

On the other hand, **Figure 5(b)** demonstrates the temperature trajectories improved by the optimal set-point control in the operational mode with the cooperative action of decentralized agents to meet the high-quality requirement.

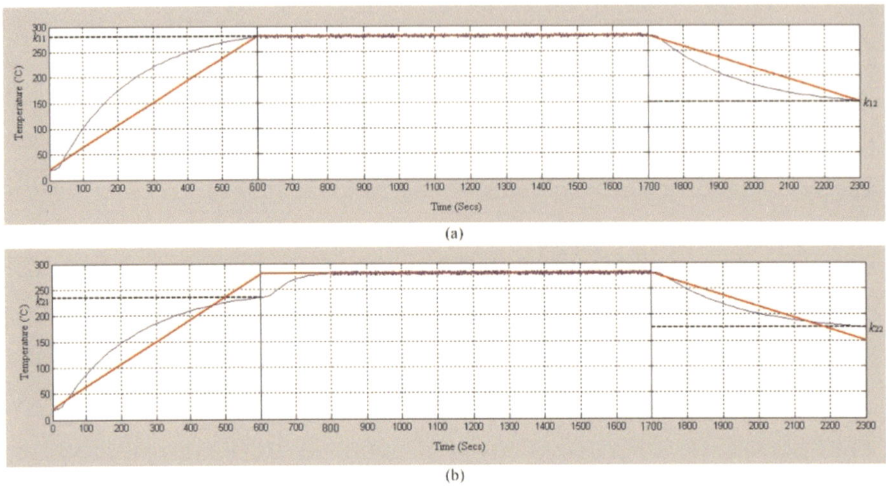

Figure 5.
Comparison of temperature change trajectory approximation. (a) Previous method. (b) Improved method.

Moreover, in the case of batch production for a limited amount of specific product, it is possible to satisfy the more accurate temperature trajectory by stepwise change of the set-point.

Experimental results show that the cooperative action of agents for determining the set-point of the decentralized local controllers ensures the economic optimization of the process operation.

6. Conclusions

This paper proposes an NMPC design problem with the hierarchical structure as an online set-point optimizer to maximize the economic efficiency of long-term operation in order to meet the strict quality requirements of products in the cascaded three-step kiln.

Firstly, a three-layer ISCS was designed, which involves a direct control layer for the kiln process, an online set-point optimization layer for decentralized controllers and a management layer for the economic optimization of the process.

Second, a distributed MPC system that approximates the temperature trajectory of the raw material of passing through the kiln to an integrated reference trajectory is proposed, which adopts the multi-agent Q-learning to optimize the set-point for decentralized controllers.

Third, a new method for fast Q-learning convergence is proposed. For the agent of the heating zone, the human operator's experience rules were automatically extracted from the process database using C4.5, and the Q-table initialization algorithm based on experience rules was developed.

In addition, for agents of the preheating and reducing zones, the Q-table initialization method using the model is proposed.

Thus, the cooperative action of multi-agent Q-learning-based ISCS determines the optimal set-points of the controllers to stabilize the process, so that it can replace the human operator in the kiln process which does not permit the trial-and-error operations. Moreover, the experiments show that fast Q-learning convergence is guaranteed by the proposed method.

This method has great significance in designing the supervisory control system for complex processes such as the ceramics, the foodstuff and the paper industry. Our future work considers the optimization problems at the plant management layer of the hierarchical MPC systems.

A. Agent design

A.1 State space

If the temperature trajectory in the preheating and reducing zones of **Figure 2** are satisfying the reference trajectory, the relation between the material quantity d, the speed v, the heating zone temperature T_2 and the product color C_p can be represented as follows:

$$(d, v, T_2) \rightarrow C_p \tag{17}$$

where v and T_2 are measured by the controllers (PLCs) and d and C_p are calculated as the luminance through the estimator from the CCD camera image. In order to represent the grey-scale information by the luminance Y from the 8-bit RGB color model, the YIQ color model [19] is used as follows:

$$Y = 0.299R + 0.587G + 0.114B, Y \in [0, 255] \tag{18}$$

d is given as the percentage of the white-colored- material area to the black-colored-belt area in the unit area from the two-value black-white image, which is converted from the grey-scale image. The state variables d, v and T_2 are discretized so that Q-learning agent could determine the optimal set-point for the speed and temperature by the iterative algorithm to meet the luminance requirement of product under the variation of the material quantity.

The state space is partitioned into fuzzy subspaces represented by five linguistic variables, that is (NB, NM, ZO, PM, PB), which can be represented from Eq. (17) as follows:

$$S = \{s | s_{ijk} = (d_i, v_j, T_{2,k}), i = 1, 2, \dots, I, j = 1, 2, \dots, J, k = 1, 2, \dots, K\} \tag{19}$$

where I, J and K are the numbers of equally divided fuzzy subsets in the proper ranges of d_i, v_j and $T_{2,k}$, respectively, $I = J = K = 5$ and the number of possible states are $5^3 = 125$.

A.2 Action space

In **Figure 2**, the set-point change, $\Delta v \in [-0.02, +0.02]$ and $\Delta T_2 \in [-15, +15]$ (°C), are the action variables of the agent for the preheating zone and the action space is partitioned into the fuzzy sub-spaces denoted by five linguistic variables as follows:

$$A = \{a | a_{lm} = (\Delta v_l, \Delta T_{2,m}), l = m = 1, 2, \dots, 5\} \tag{20}$$

The number of possible actions for each state (d, v, T_2) is $5^2 = 25$. The agent chooses a pair of state-action from Q-table based on the observed states and the ε-greedy policy of choosing actions.

According to this, the rules for updating set-points for the speed and the temperature are as follows:

$$v(t+1) = v(t) + \Delta v(t+1) \tag{21}$$

$$T_2(t+1) = T_2(t) + \Delta T_2(t+1) \tag{22}$$

where $v(t+1)$ and $T_2(t+1)$ are the speed and temperature set-points in the present time, $v(t)$ and $T_2(t)$ are ones in the past time, respectively.

A.3 Reward function

Since the product color C_p as the quality characteristic depends on the d, v and T_2, the luminance is used as the reward of the reinforcement learning for the heating zone.

Due to the action of the agent, the state s_t proceeds to s_{t+1} and the product color $C_p(t)$ changes to $C_p(t+1)$. The reward r_{t+1} given to the agent is as follows:

$$r_{t+1} = f\left(C_p(t+1)\right) \tag{23}$$

Considering the experimental results, the luminance value of Eq. (18) is partitioned into five fuzzy sets, as shown in **Figure 6**, and Eq. (23) is defined as the discrete reward function such as Eq. (24).where the value of Y' is the fuzzy variable for the linguistic evaluation of the luminance Y.

$$r_{t+1} = \{2, Y' = \text{ZO}|\{1, Y' = \text{PM}|\{0, Y' = \text{NM}|\{-1, Y' = \text{PB}|\{-2, Y' = \text{NB}| \tag{24}$$

B. Convergence improvement method of Q-learning

The scheme of improving the Q-learning convergence for the heating zone agent involves procedures such as (i) Extracting the decision tree for the operational

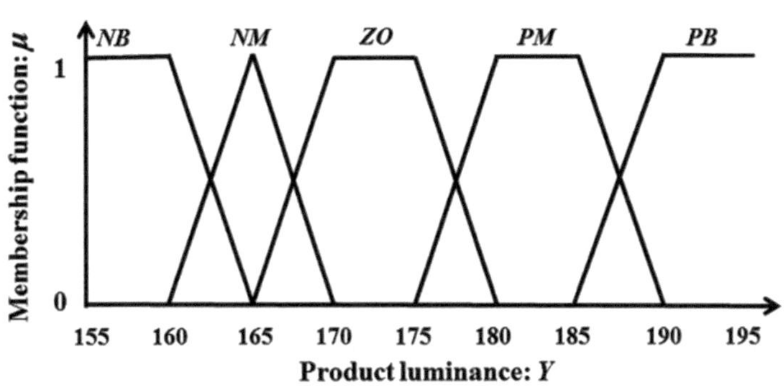

Figure 6.
Fuzzy set of luminance value.

experience rules from the historical database, (ii) Converting the decision tree to fuzzy rule, (iii) Initializing the Q-table by the fuzzy rule.

B.1 Extraction of operating rules from data

B.1.1 Extraction of instances set

The data records $R(At_1, At_2, \ldots, At_k, \ldots)$ stored in the database involves the state variable set $X = \{x_1, x_2, \ldots, x_k\}$, the action variable set (set-points) $Y = \{y_1, y_2, \ldots, y_p\}$, the operator's experience knowledge and the process dynamics.

Thus, the new relationship between the state and action is obtained using the relational projection operation [20] as follows:

$$\prod_q (R) = \{t[q] \mid t \in R\} \tag{25}$$

where $t[q]$ is the set of the value of the attribute q to the element t of the relation R.

First, samples are selected from the created relations and each element of the sample is normalized onto the interval $[0, 1]$. And the normalized samples are classified according to the Euclidean similarity criteria of Eq. (26) and the obtained instances are classified randomly into training data and test data:

$$r_{\text{ij}} = \left[\sum_{k=1}^{m} \left(q_{\text{ik}} - q_{\text{jk}}\right)^2 / m\right]^{1/2} \tag{26}$$

where $q_i = (q_{i1}, q_{i2}, \ldots, q_{im})$ and $q_j = (q_{j1}, q_{j2}, \ldots, q_{jm})$ are ith and jth samples.

The obtained instance set D consists of two classes according to the value C of class that if the luminance which shows the product quality lies in the proper range δ then $C = 1$ else $C = 0$, and its structure is shown in **Table 3**. Then, $D_1 \cap D_0 = \phi$.

In **Table 3**, the attribute set $I = \{I_1, I_2, \ldots, I_m\}$ contains the state variable set X and the action variable set Y, where $I = X \cup Y$, $X \cap Y = \phi$ and $m = k + p$.

B.1.2 Rule generation

The decision tree learning method to extract the operating rules from data utilizes the C4.5 algorithm [21], which is the successor of ID3 [22].

	I_1	I_2	I_3	…	I_m	Class
1	x_{11}	x_{12}	x_{13}		x_{1m}	C_1
2	x_{21}	x_{22}	x_{23}		x_{2m}	C_2
3	x_{31}	x_{32}	x_{33}		x_{3m}	C_3
…	…	…	…	…	…	…
n	x_{n1}	x_{n2}	x_{n3}		x_{nm}	C_n

Table 3.
Structure of instance set.

C4.5 algorithm for constructing the decision tree includes the following contents:

i. To choose the root node.

ii. To construct the branches based on every value of the root node.

iii. To continue the construction of the decision tree until no further disintegration occurs to the attributes.

The fuzzification of attributes is conducted to extract the fuzzy rules such as Eq. (27) from **Table 3**.

$$\text{Rule } R_j : \text{If } x_1 \text{ is } A_{j1} \text{ and} \dots \text{and } x_m \text{ is } A_{jm} \text{ then class } C_j, j = 1, 2, \dots, N \tag{27}$$

where $\mathbf{X} = [x_1, x_2, \dots, x_m]$ is the m-dimensional sample vector, A_{ji} $(i = 1, 2, \dots, m)$ is the fuzzy set of x_j; $C_j \in \{0, 1\}$ is the pattern class of the jth rule and N is the number of fuzzy rules.

In the decision tree constructed from the training data by the C4.5 algorithm, each path from a root to a leaf represents a rule.

Table 4 shows some rules obtained from the decision tree.

B.1.3 Obtaining of state-action rules

The state-action pairs with the C = 1 in **Table 4** represent the operating conditions applicable to the process. Therefore, by choosing only the rules with C = 1 among the N rules, the fuzzy rules for the following state-action pairs would be obtained:

$$\text{Rule } R_k : \text{If } x_1 \text{ is NM and } x_2 \text{ is NM and } x_3 \text{ is NM then } y_1 \text{ is NM and } y_2 \text{ is PM} \tag{28}$$

where $k = 1, 2, \dots, M$ is the number of rules and $M < N$.

B.2 Initialization method of Q-table

From Eq. (28), as each rule has 3 fuzzy variables for state and 2 fuzzy variables for action, the Q-table contains 125 states, each of which has 25 actions.

However, since only the empirical knowledge of operator is stored in the process database, the fuzzy rule obtained from **Table 4** reveals only the most appropriate action to the given state among the possible action space.

Rule	Description
R1	x_1 = NB & x_2 = PM & x_3 = ZO & y_1 = PB & y_2 ZO ⇨ C = 1
R2	x_1 = PM & x_2 = PM & x_3 = ZO & y_1 = ZO & y_2 = PM ⇨ C = 1
R3	x_1 = ZO & x_2 = PM & x_3 = ZO & y_1 = NB & y_2 = PM ⇨ C = 0
R4	x_1 = NB & x_2 = PM & x_3 = ZO & y_1 = NB & y_2 = PM ⇨ C = 0
R5	x_1 = ZO & x_2 = ZO & x_3 = ZO & y_1 = ZO & y_2 = ZO ⇨ C = 1

Table 4.
Some of the decision tree rules.

1. $i = 1$.
2. Search the rule of Eq. (28), whose antecedent matches the ith state s_i in Q-table. If the rule does not exist, then go to step 5.
3. Assign maximum value to the Q-value of an action that matches with the consequence of the rule.
4. Initialize Q-values of the rest actions using Eq. (31) and go to step 6.
5. Initialize Q-values of all possible actions to the state s_i as zero.
6. $i \leftarrow i + 1$.
7. Repeat from step 2 for all the states.

Table 5.
Initialization algorithm of Q-table.

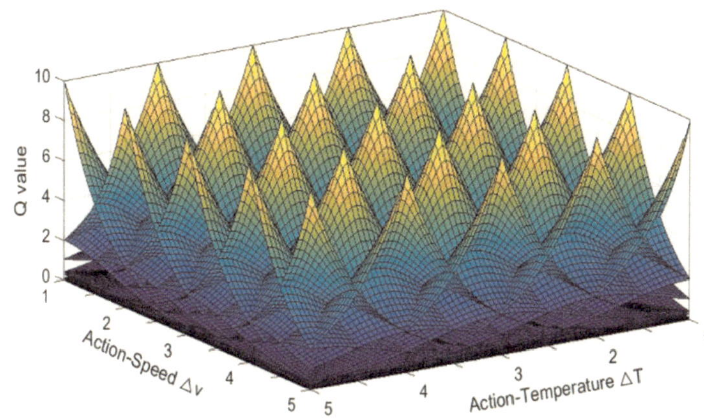

Figure 7.
Assignment of Q-value in 5 × 5 action space.

Thus, we propose the algorithm of initializing the Q-value for the action space that cannot be obtained by the decision tree for each state in **Table 5**.

The exponential function distribution used in the determination of Q-values in the three-dimensional space is represented as follows:

$$Q_0(w_1, w_2) = A \cdot \exp(-k\|\mathrm{w} - \mathrm{r}\|) \tag{29}$$

where ||**w-r**|| is the Euclidean Square Norm of feature vectors defined as follows:

$$\|\mathrm{w} - \mathrm{r}\| = \left(\sum_{l}^{n}\left(|w_l - r_l|^2\right)\right)^{1/2} \tag{30}$$

Since $n = 2$, the function for initializing Q-value is represented as follows:

$$Q_0(w_1, w_2) = A \cdot \exp\left(-k\left((w_1 - r_{1i})^2 + (w_2 - r_{2j})^2\right)^{1/2}\right), (w_1, w_2) \in W \tag{31}$$

where actions w_1 and w_2 have the values of Δv and ΔT, respectively, W is the action set and $i = j = 1, 2, \ldots, 5$. r_{1i} and r_{2j} are respectively the centres of the ith and jth action range. A is the initial constant value and k is the growth constant. In this study, $A = 10$ and $k = 1.6$, respectively.

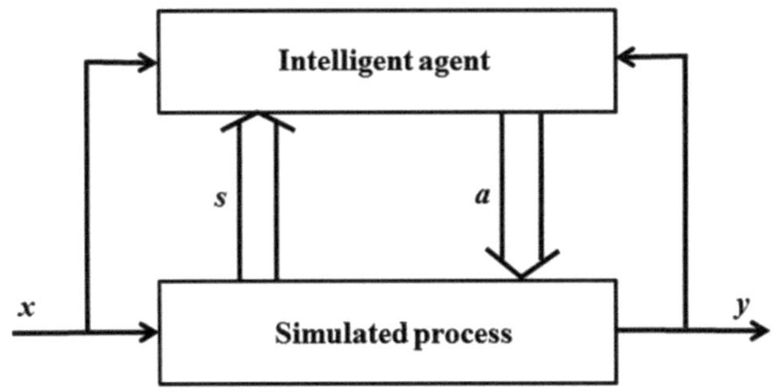

Figure 8.
Architecture of simulation system.

The function plotted in MATLAB is shown in **Figure 7**.

C. Simulation results

Figure 8 shows the architecture of the simulation system, which is used to evaluate the performance of the agent for the heating zone under the assumption that the direct control layer in **Figure 2** performs well enough.

The reward for the agent is calculated by Eq. (24) and the reward signal is simulated by the following luminance prediction model corresponding to the color of product obtained using the process data:

$$y = 170.34 + 5.97x_1 + 2.87x_2 - 0.894x_3 \tag{32}$$

where x_1, x_2, x_3 and y denote the quantity of material d, the speed v, the temperature T and the product luminance Y, respectively. Thus, the agent observes the states of the process through x_1, x_2 and x_3, and takes the actions through Δx_2, Δx_3, and estimates the reward by y. During simulation, the Q-table and parameters for the real process are utilized (**Table 6**).

The simulation and experiment results are shown in **Figures 9–11**.

Figure 9 compares the Q-learning convergence of the proposed method to the traditional one in regard to the Bellman error [23].

Figure 10 demonstrates the operating characteristic during simulation.

Parameter	Variable	Value	unit
Learning rate	α	0.1	
Discount factor	γ	0.9	
Material quantity	d	$10 \sim 50$	%
Belt speed	v	$3 \sim 7$	m/min
Temperature	T	$180 \sim 340$	°C
Luminance	C_p	$155 \sim 195$	

Table 6.
Parameters for simulation.

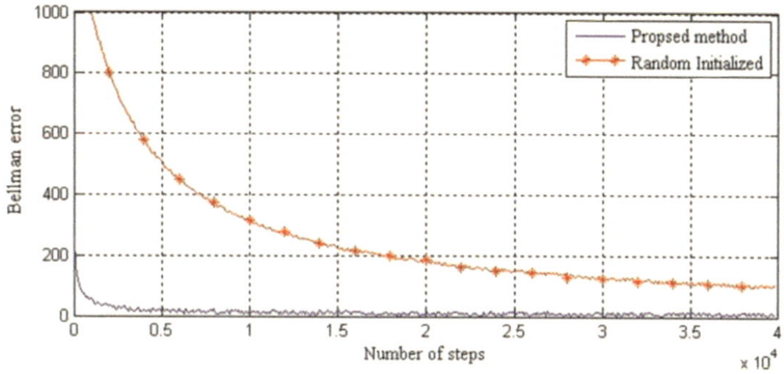

Figure 9.
Comparison of Q-learning convergence.

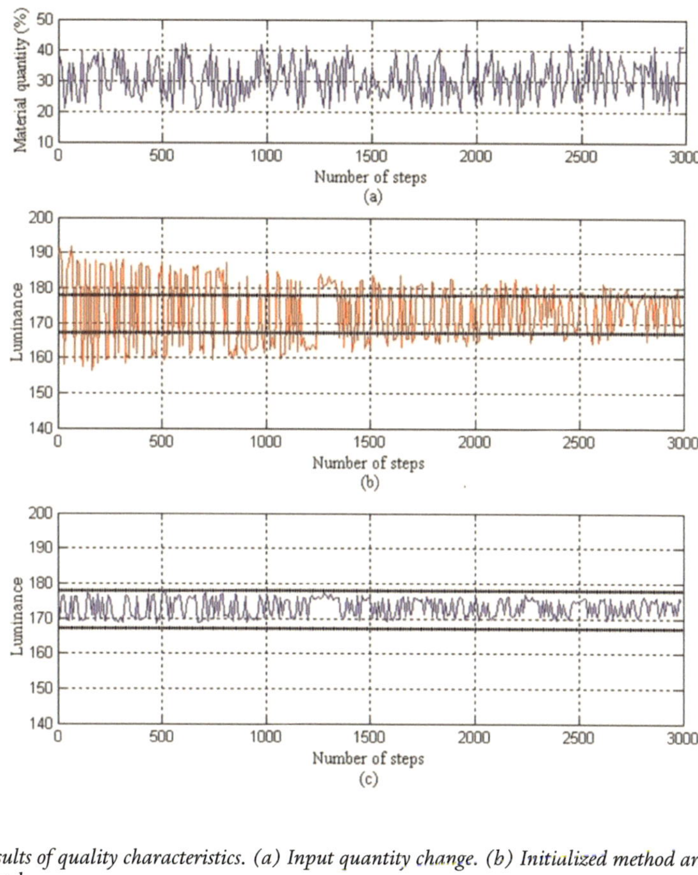

Figure 10.
Simulation results of quality characteristics. (a) Input quantity change. (b) Initialized method arbitrarily. (c) Proposed method.

In **Figure 10(a)**, the virtual input x_1 changes randomly and the variables x_2, x_3 are given as certain values according to the process.

In **Figure 10(b)** and **(c)**, the upper and lower limit of the luminance corresponding to the product quality is shown as the thick line.

The luminance is calculated by Eq. (32) and the reward is predicted by Eq. (24).

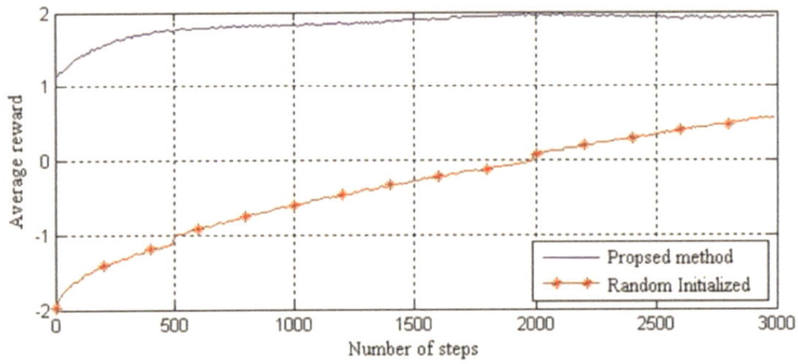

Figure 11.
Comparison of average reward.

The deviation of luminance is V = 5.66% for the traditional method, whereas V = 0 during learning iterations for the proposed method.

Figure 11 shows the comparative result of the proposed method with the random initialized one by the average reward method [14].

Simulation results show that the proposed method satisfies the Q-learning convergence and the function optimality in the initial learning phases.

Author details

Song Ho Kim* and Kwang Rim Song
Faculty of Automatics, Kim Chaek University of Technology, Pyongyang, Democratic People's Republic of Korea

*Address all correspondence to: ksh6436@star-co.net.kp

References

[1] Fekri S, Assadian F. Fast model predictive control and its application to energy Management of Hybrid Electric Vehicles. In: Zheng T, editor. Advanced Model Predictive Control. London, UK: IntechOpen; 2011. pp. 3-28 ch1

[2] Ocampo-Martinez C. Model Predictive Control of Wastewater Systems. Springer-Verlag London Limited; 2010. p. 216. DOI: 10.1007/978-1-84996-353-4

[3] Wang L. Model Predictive Control System Design and Implementation Using MATLAB®. Springer-Verlag London Limited; 2009. p. 374. DOI: 10.1007/978-1-84882-331-0

[4] Qin SJ, Badgwell TA. A survey of industrial model predictive control technology. Control Engineering Practice. 2003;**11**(7):733-764

[5] Arellano-Garcia H, Barz T, Dorneanu B, Vassiliadis VS. Real-time feasibility of nonlinear model predictive control for semi-batch reactors subject to uncertainty and disturbances. Computers and Chemical Engineering. 2020;**133**:1-18. DOI: 10.1016/j.compchemeng.2019.106529

[6] Patil BV et al. Decentralized nonlinear model predictive control of a multi-machine power system. Electrical Power and Energy Systems. 2019;**106**:358-372. DOI: 10.1016/j.ijepes.2018.10.018

[7] Zhang A, Yin X, Liu S, Zeng J, Liu J. Distributed economic model predictive control of wastewater treatment plants. Chemical Engineering Research and Design. 2019;**141**:144-155. DOI: 10.1016/j.cherd.2018.10.039

[8] Brdys M, Tatjewski P. Iterative Algorithms for Multilayer Optimizing Control. London, UK: Imperial College Press; 2005. p. 370

[9] Marusak P, Tatjewski P. Actuator fault toleration in control systems with predictive constrained set-point optimizers. International Journal of Applied Mathematics and Computer Science. 2008;**18**(4):539-551. DOI: 10.2478/v10006-008-0047-2

[10] Tatjewski P. Supervisory predictive control and on-line set-point optimization. International Journal of Applied Mathematics and Computer Science. 2010;**20**(3):483-495. DOI: 10.2478/v10006-010-0035-1

[11] Li H, Swartz CLE. Dynamic real-time optimization of distributed MPC systems using rigorous closed-loop prediction. Computers and Chemical Engineering. 2018;**122**:356-371

[12] Kim SH, Song KR, Kang IY, Hyon CI. On-line set-point optimization for intelligent supervisory control and improvement of Q-learning convergence. Control Engineering Practice. 2021;**114**:104859. DOI: 10.1016/j.conengprac.2021.104859

[13] Watkins CJCH, Dayan P. Technical note: Q-learning. Machine Learning. 1992;**8**:279-292

[14] Montague PR. Reinforcement Learning: An Introduction, by Sutton RS, Barto AG. Trends in Cognitive Science 3, no 9; 1999. p. 360

[15] Zarrabian S, Belkacemi R, Babalola AA. Reinforcement learning approach for congestion management and cascading failure prevention with experimental application. Electric Power Systems Research. 2016;**141**:179-190. DOI: 10.1016/j.epsr.2016.06.041

[16] Chen X, Chen G, Cao W, Wu M. Cooperative learning with joint state value approximation for multi-agent systems. Journal of Control Theory and Applications. 2013;**11**(2):149-155. DOI: 10.1007/s11768-013-1141-z

[17] Wang H, Wang X, Hu X, Zhang X, Gu M. A multi-agent reinforcement learning approach to dynamic service composition. Information Sciences. 2016;**363**:96-119. DOI: 10.1016/j.ins.2016.05.002

[18] Zemzem W, Tagina M. Cooperative multi-agent systems using distributed reinforcement learning techniques. Procedia Computer Science. 2018;**126**: 517-526. DOI: 10.1016/j.procs.2018.07.286

[19] Girgis MR, Mahmoud TM, Abd-El-Hafeez T. An approach to image extraction and accurate skin detection from web pages. International Journal of Electrical and Electronics Engineering. 2007;**1**(6):353-361

[20] Codd EF. The Relational Model for Database Management: Version 2. Menlo Park: Addison-Wesley Publishing Company; 1990. p. 538

[21] Quinlan JR. C4.5: Programs for Machine Learning. San Francisco: Morgan Kaufmann Publishers; 1993

[22] Quinlan JR. Induction of decision trees. Machine Learning. 1986;**1**(1): 81-106

[23] Buşoniu L, Babuška R, Schutter B, Ernst D. Reinforcement Learning and Dynamic Programming Using Function Approximators. Boca Raton: CRC Press; 2010. p. 267

Chapter 5

Closed Loop Control Approaches for Conflicted Timed Event Graphs Subject to Constraints with Dioid Algebra

Nesrine Ben Afia and Hassani Messaoud

Abstract

In this chapter, we are going to tackle the control of a particular class of Discreet Event systems, which is the subclass of Conflicted Timed Event Graphs with the use of Dioid Algebra (Min, +). The main proposed approaches concern systems including sharing phenomena and are subject to various constraints such as time, capacity, and mixed constraints. For that, we establish a convenient switching policy according to a specific allocation order, and we compute suitable control laws that guarantee the proper system functioning while ensuring the respect of its constraints.

Keywords: conflicting timed event graphs, constraints, resource sharing, switching models, control laws, Dioid algebra (Min, +)

1. Introduction

For Discrete Event Systems, resource sharing could be a problem. Therefore, time and space compliance is crucial, and a loss of time and space results when the right control decision is not made, especially when various constraints exist. For this reason, the allocation of shared resources for Timed Event Graphs (TEGs) between several processes can be transformed into a switching system control problem that is modeled using the (Min, +) algebra. As this type of system could contain a finite number of states, then it belongs consequently to the class of Discrete Event Systems (DES), the object of our study.

This chapter book takes place among the investigations and the studies in order to establish control approaches for Discrete Event Systems [1–4]. In several industrial processes, both time and capacity factors are critical conditions that affect the evolution of Discrete Events, especially when resource-sharing phenomenons arise, which urges us to define a suitable modeling adapted to their nature and following switching operational modes.

To do this, we have placed ourselves in relation to the works done with the use of the same tools in order to establish an adapted state of the art to our purposes. We have been able to improve some approaches by relaxing some limiting assumptions, and the new approaches established preserve the liveliness of Timed Event Graphs by avoiding dead loops without tokens [5, 6]. An effective switching policy is introduced to organize the access to the shared resource. Switching (Min, +) models are then adapted, computed, and used to solve the control problem under the existence of constraints. We have established an adapted control policy according to a sequence of active modes of competing graphs and defined according to an operating plan. For our case study, we were interested not only in mixed modes but also in conservative modes for which the same graph within a mode can exclusively appropriate the resource allocation.

Original (Min, +) control synthesis is detailed in the case of the existence of a single marking constraint on one of the CTEG's places, which is after that generalized to the case of multiple marking constraints, and we have also dealt with the problem of possible mixed constraints of both time and capacity. The design of control laws that satisfy these requirements is therefore an important step, and sufficient conditions for the satisfaction of these constraints have been deduced.

2. Modeling of CTEGs

To model the sharing phenomenon between CTEGs, we first study the switching behavior of these systems. Their evolution is described by the succession of modes, which constitutes a switching sequence defining the resource allocation of a particular process, whereby the organization of modes within a sequence is already fixed by the production plan. In the following section, we give some basic definitions.

2.1 Definitions

A Conflicted Timed Event Graph is defined by $N \in \mathbb{N}^* \setminus \{1\}$ Timed Event Graphs of type SISO.

A mode is a selection of a number of N Timed Event Graphs competing to win the allocation of the shared resource.

A sequence, denoted by S, is a set of modes whose order is fixed according to a particular operational need.

Remark 3.1: In our study, the possession of the resource token can be considered by the same Timed Event Graph in a mode. This means that we are in a conservative mode. In the case where different Timed Event Graphs are in competition within the same mode, it is called a mixed mode.

2.2 Hypotheses

- N defines the number of concurrent SISO Timed Event Graphs (users) such that $N > 1$.
- Taking as K the number of shared resources, in order to respect the competition criteria, we consider the following condition $K < N$.
- Since the periodic allocation of the shared conflict place depends on the order of the modes in the sequence, it results in a periodic Conflicted Timed Event Graph.

- The token in the conflict place is consumed only once by a single graph. In terms of Petri nets when this place is not marked, it means that there is only one active elementary circuit in possession of the resource.
- We consider a maximum transition firing speed rate.
- For each internal transition t_i of the CTEG, a counter function noted θ_i is associated, and for each source transition noted t_{u_i}, a resource transition noted u_i^k is associated, where K is the mode index.

2.3 (Min, +) switching model of a CTEG

Most of the works dealing with control of Petri nets, under marking constraints, have not taken into account the time parameter in the modeling phase or in the control synthesis phase. Despite the fundamental role of the time factor, a few studies have dealt with the control of SEDs modeled by timed Petri nets.

For the example of the TEG shown in **Figure 1**, we observe that three Timed Event Graphs (users) compete to win the resource allocation in one mode. Furthermore, we provide the rules for managing the switching between the different modes to validate the allocation of the shared resource. As defined earlier, a mode can contain a number of competing graphs. The activation logic of the transitions follows the order of the selected graphs, such that only the upstream and downstream transitions of the shared resource are concerned on the modeling step.

We highlight the modeling step of the GETC, especially at the shared resource level; see **Figure 1**. For this, we chose the operational mode $M_k(G_n G_{n+1})$ where k is the mode index. Thus, the activation of the mode systematically leads to the possession of the token by the first graph noted G_n (respectively with its upstream and downstream transitions linked to the shared resource). After having been used for a certain time τ, the resource will be available for the graph G_{n+1}, and the token will thus be distributed by the output transition t_k to the input transition t_i of this mode

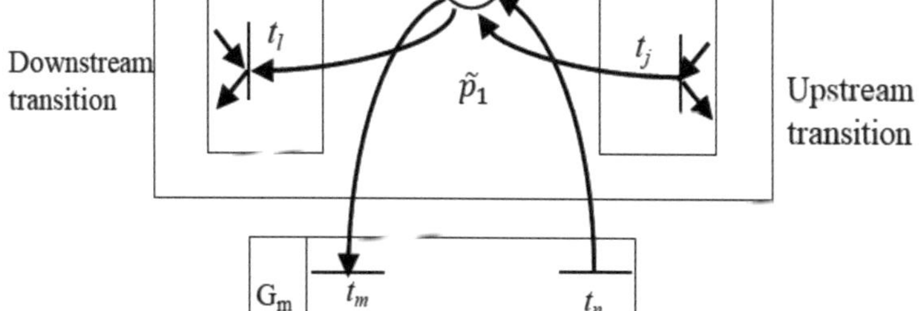

Figure 1.
Highlighting the conflict at the level of the shared resource $\tilde{p}_1$.

(the other arcs connected to the resource with other transitions will not be considered by the modeling step).

Referring to [7] and by analogy to TEGs, the modeling of GETC is based on the swap/switch between TEGs under the switching model approach.

Remark 1: Each defined mode is associated with a switching model (Min, +). The dynamic evolution equation corresponding to the activation of the mode is noted by M_k in the sequence S, and it is defined by the following equation:

$$x_x(t) = A_x.x(t-1) \oplus B_x u_x(t) \quad (1)$$

Where x represents the index of the corresponding switching model (Min, +), we have:

$A_x \in \mathbb{R}_{\min}^{n*n}$ is the state matrix of the mode k.

$B_x \in \mathbb{R}_{\min}^{n*m}$ is the control matrix of the mode k.

u_x is the control vector of the mode k.

n is the number of all internal transitions, and m is the number of source transitions.

Example: In **Figure 2**, the CTEG is composed of 3 Timed Event Graphs respectively given by: GET 1, GET 2, and GET 3. The resource allocation given by the sequence S as follows:

$$S = M_1 M_2 \quad (2)$$

Considering a maximum firing speed, the first mode M_1 is defined by $M_1(G_1 G_3)$; the resource occupation of $\tilde{p}_1$ is interpreted by a token moving for the activation of the graph G_1 then the graph G_3. The counter functions associated with the source transitions t_{u_1} and t_{u_2} are respectively u_1^1 and u_3^1. θ_i indicates the counter function that corresponds to the CTEG transitions t_i, such that $i = 1, \ldots, 8$.

The modeling steps consist of determining the equations of the system constituted by counter functions associated to each transition.

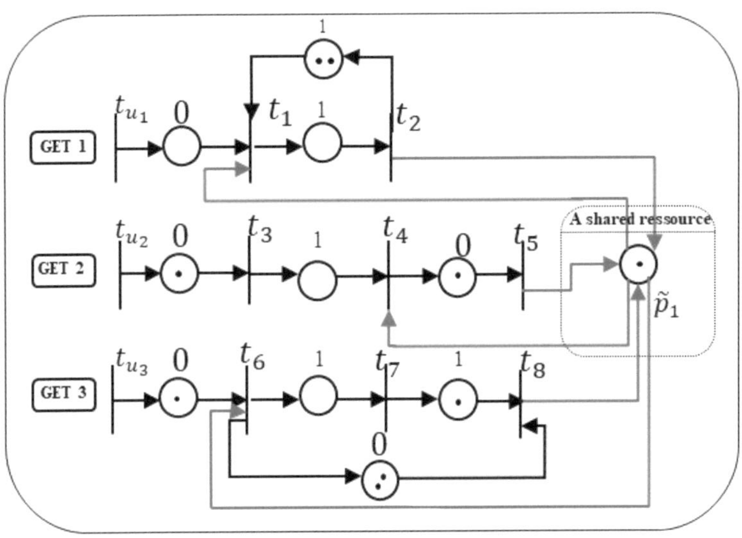

Figure 2.
An example of three TEGs (TEG1, TEG2, and TEG3) sharing a conflict place $\tilde{p}_1$.

To obtain the linear (Min, +) model and for more clarity sakes, it is recommended to make these changes; we replace "Min" by "⊕" and "+" by ".", and one obtains, consequently, the following system of (Min, +) linear equations:

$$\begin{cases} \theta_1(t) = e.u_1^1(t) \oplus 2.\theta_2(t-1) \\ \theta_2(t) = e.\theta_1(t-1) \\ \theta_3(t) = 1.u_2^1(t) \\ \theta_4(t) = e.\theta_3(t-1) \oplus 1.\theta_8(t) \\ \theta_5(t) = 1.\theta_4(t) \\ \theta_6(t) = e.u_3^1(t) \\ \theta_7(t) = e.\theta_6(t-1) \\ \theta_8(t) = 1.\theta_7(t-1) \end{cases} \tag{3}$$

When the system switches the mode from M_1 to M_2, the model also changes. For the second mode $M_2(G_3G_2)$, we obtain the following system of (Min, +) equations:

$$\begin{cases} \theta_1(t) = e.u_1^2(t) \oplus 2.\theta_2(t-1) \\ \theta_2(t) = e.\theta_1(t-1) \\ \theta_3(t) = 1.u_2^2(t) \\ \theta_4(t) = e.\theta_3(t-1) \oplus 1.\theta_8(t) \\ \theta_5(t) = 1.\theta_4(t) \\ \theta_6(t) = e.u_3^2(t) \\ \theta_7(t) = e.\theta_6(t-1) \\ \theta_8(t) = 1.\theta_7(t-1) \end{cases} \tag{4}$$

After determining the switching models (Min, +) corresponding to these modes, we provide the state space representations associated with each one as follows:

For $M_1(G_1G_3)$, we have:

$$x_1(t) = \begin{bmatrix} \varepsilon & \varepsilon & 2 & \varepsilon & \varepsilon & \varepsilon & \varepsilon & \varepsilon \\ e & \varepsilon & \varepsilon & \varepsilon & \varepsilon & \varepsilon & \varepsilon & \varepsilon \\ \varepsilon & 1 & \varepsilon & \varepsilon & \varepsilon & \varepsilon & \varepsilon & \varepsilon \\ \varepsilon & \varepsilon & e & \varepsilon & \varepsilon & \varepsilon & \varepsilon & \varepsilon \\ \varepsilon & \varepsilon & 2 & \varepsilon & \varepsilon & \varepsilon & \varepsilon & \varepsilon \\ \varepsilon & \varepsilon & \varepsilon & \varepsilon & \varepsilon & \varepsilon & \varepsilon & \varepsilon \\ \varepsilon & \varepsilon & \varepsilon & \varepsilon & \varepsilon & \varepsilon & \varepsilon & \varepsilon \\ \varepsilon & \varepsilon & \varepsilon & \varepsilon & \varepsilon & e & \varepsilon & \varepsilon \end{bmatrix} .x(t-1) \oplus \begin{bmatrix} e & \varepsilon & \varepsilon \\ \varepsilon & \varepsilon & \varepsilon \\ \varepsilon & 1 & \varepsilon \\ \varepsilon & \varepsilon & \varepsilon \\ \varepsilon & \varepsilon & \varepsilon \\ \varepsilon & \varepsilon & e \\ \varepsilon & \varepsilon & \varepsilon \\ \varepsilon & \varepsilon & \varepsilon \end{bmatrix} .u(t) \tag{5}$$

And for the second mode $M_2(G_2G_3)$, the following state space equation is achieved:

$$x_2(t) = \begin{bmatrix} \varepsilon & 2 & \varepsilon & \varepsilon & \varepsilon & \varepsilon & \varepsilon & \varepsilon \\ e & \varepsilon & \varepsilon & \varepsilon & \varepsilon & \varepsilon & \varepsilon & \varepsilon \\ \varepsilon & \varepsilon & \varepsilon & \varepsilon & \varepsilon & \varepsilon & \varepsilon & \varepsilon \\ \varepsilon & \varepsilon & e & \varepsilon & \varepsilon & \varepsilon & 2 & \varepsilon \\ \varepsilon & \varepsilon & 1 & \varepsilon & \varepsilon & \varepsilon & \varepsilon & \varepsilon \\ \varepsilon & \varepsilon & \varepsilon & \varepsilon & \varepsilon & \varepsilon & \varepsilon & \varepsilon \\ \varepsilon & \varepsilon & \varepsilon & \varepsilon & \varepsilon & e & \varepsilon & \varepsilon \\ \varepsilon & \varepsilon & \varepsilon & \varepsilon & \varepsilon & \varepsilon & 1 & \varepsilon \end{bmatrix} .x(t-1) \oplus \begin{bmatrix} e & \varepsilon & \varepsilon \\ \varepsilon & \varepsilon & \varepsilon \\ \varepsilon & \varepsilon & \varepsilon \\ \varepsilon & \varepsilon & \varepsilon \\ \varepsilon & \varepsilon & \varepsilon \\ \varepsilon & \varepsilon & \varepsilon \\ \varepsilon & 1 & \varepsilon \\ \varepsilon & \varepsilon & \varepsilon \end{bmatrix} u(t) \tag{6}$$

3. Formulation of the control problem

Space constraints are crucial factors that are often considered in the context of the Petri net formalism. Our main objective is to find efficient controllers to optimize this constraint by limiting the number of tokens in certain places and that is depending on the system needs. These places could be assimilated to stations in transportation networks or to limited stock areas in industrial contexts. The marking constraints are expressed by linear inequalities.

In the same context, among some recent contributions, we can mention [8], where the authors addressed the control design problem to synthesize a causal control feedback that serves to satisfy the capacity constraints in a single place as well as in a sequence of places. In this sense, they tried to establish relaxed adaptive assumptions by assuming that each linear marking constraint only affects a single place or places located on the same path within the TEG.

Through this section, we will give some hypotheses in order to solve the problem of undesirable situations related to capacity constraints (whatever the place) for a large and complex case of Timed Petri nets, which is Conflicting Timed Event Graphs. To do so, it is necessary to explain some interesting concepts.

In the following sections, after introducing the marking constraint, we will proceed first by the control synthesis formulation for the case of a single constraint before extending to the case of multiple constraints.

3.1 Marking constraint

An accessible marking vector represents the dynamical behavior of Petri nets. Forbidden states are markings that can be determined, but they are at the same time still forbidden by specifications imposed to the system.

Since capacity is not only a cost-of-space issue but also a price issue, we spend a lot of time finding control law that respects this critical constraint.

For a TEG modeled through the DES class, we assume that a place p_{ij} is exposed to a capacity constraint. Thus, regardless of context necessity, there is a finite number of tokens that must not be exceeded.

For the place p_{ij}, $M_{ij}(t)$ denotes the marking of this place until t time, which is expressed by the following inequality:

$$M_{ij}(t) = x_j(t) - x_i(t) + M_{ij0} \tag{7}$$

Where $x_i(t)$ and $x_j(t)$ are, respectively, the counter functions of the input and output transitions related to this place. The first counter function $x_i(t)$ represents the number of transitions firing from the transition start time t_j to t time. Thus, $x_j(t)$ defines the sum of input tokens, and $x_i(t)$ defines the sum of output tokens.

In the case of marking constraint, the final marking of this place from an initial marking M_{ij0} must not exceed a certain threshold noted b. Eq. (7) can thus be reduced through (Min, +) algebra to the following inequality, which represents the marking constraint that we seek to satisfy:

$$x_j(t) \leq \left(b - M_{ij0}\right).x_i(t) \tag{8}$$

Given the explicit equation of state representation: $x(t) = A.x(t-1) \oplus B.u(t)$ such that we have $A = \hat{A}_0^*.\hat{A}_1$ and $B = \hat{A}_0^*.\hat{B}$, its equivalent equation is given by the following equation:

$$x(t) = A^{\tau}.x(t-\tau) \oplus \left[\bigoplus_{k=0}^{\tau-1} A^k.B.u(t-k)\right], \text{Such as } \tau \geq 1 \tag{9}$$

If we substitute $\tau = \phi$ in Eq. (9), we could establish the following explicit expression:

$$x_i(t) = \left[\bigoplus_{r=1}^{n} \left(A^{\phi}\right)_{ir} x_r(t-\phi)\right] \oplus \left[\bigoplus_{k=0}^{\phi-1} \left(A^k.B\right)_i.u(t-k)\right], \phi \geq 1, \text{and } A \in \overline{\mathbb{R}}_{\min}^{n \times n} \tag{10}$$

Where n represents the number of internal transitions of CTEG, and $\left(A^{\phi}\right)_{ir}$ denotes the ith components of the matrix.

For each path in CTEG the selected mode, we define by τ_{α_n} the sum of all delays for each TEG connecting the source transition to the upstream transition of the constrained place. We denote by n the graph number.

$$x_{j_n}(t) \leq \left(A^{\tau_{\alpha_n}} B\right)_{j_n} u_x(t-\tau_{\alpha_n}), \forall N \in \mathbb{N}^* \setminus \{1\} \tag{11}$$

By replacing the two Eqs. (10) and (11) in Eq. (9), we obtain the following expression:

$$\left(A^{\tau_{\alpha_n}} B\right)_{j_n} u_x(t-\tau_{\alpha_n}) \leq \left(b - M_{ij0}\right).\left[\bigoplus_{r=1}^{n} \left(A^{\phi}\right)_{i_n r} x_r(t-\phi)\right] \oplus \left[\bigoplus_{k=0}^{\phi-1} \left(A^k.B\right)_{i_n}.u(t-k)\right] \tag{12}$$

We see that the satisfaction of the above inequality induces the simultaneous satisfaction of the following two inequalities:

$$\left(A^{\tau_{\alpha_n}} B\right)_{j_n} u_x(t-\tau_{\alpha_n}) \leq \left(b - M_{ij0}\right)\left[\bigoplus_{k=0}^{\phi-1} \left(A^k.B\right)_{i_n}.u(t-k)\right] \tag{13}$$

$$\left(A^{\tau_{\alpha_n}} B\right)_{j_n} u_x(t-\tau_{\alpha_n}) \leq \left(b - M_{ij0}\right)\left[\bigoplus_{r=1}^{n} \left(A^{\phi}\right)_{i_n r} x_r(t-\phi)\right] \tag{14}$$

These inequalities could also be formulated as follows:

$$(A^{\tau_{\alpha n}}B)_{j_n} u_x(t - \tau_{\alpha_n}) \leq (b - M_{ij0}) \left[\bigoplus_{k=0}^{\phi-1} (A^k.B)_{i_n} .u(t-k) \right] \tag{15}$$

$$u_x(t - \tau_{\alpha_n}) \leq \left(b - M_{ij0} - (A^{\tau_{\alpha n}}B)_{j_n}\right) \left[\bigoplus_{r=1}^{n} (A^{\phi})_{i_n r} x_r(t-\phi) \right] \tag{16}$$

3.2 Control synthesis in the case of a single marking constraint

In this section, we address the problem resolution of control of CTEGs, having a safe shared resource for which we will start with the case of a single marking constraint.

Theorem 1: A CTEG consisting of N TEGs, whose evolution is given by the explicit equation of state representation (1), sharing a safe resource noted $\tilde{p}_1$ and subject to a single capacity constraint of the form (8), admits a control law $u(t)$ of the form:

$$u(t) = \bigoplus_{1}^{n} F.x(t-1) \tag{17}$$

Like:
$F = \left(b - M_{ij0} - (A^{\tau_{\alpha n}}B)_{j_n}\right).A^{\phi}$ such that $\phi = \tau_{\alpha_n} + 1$; $F \geq e$.
If the following N conditions are true:

$$(A^{\tau_{\alpha n}}B)_{j_n} \leq (b - M_{ij0}).(A^k.B)_{i_n} \; k = 0, \ldots, \tau_{\alpha_n} \tag{18}$$

Proof is provided in [9].

3.3 Control synthesis for the case of multiple marking constraints

In this section, we deal with a more complex case, where multiple users are competing to win the resource allocation. In addition to this complexity criterion that could be cumulated especially with the presence of multiple marking constraints. Therefore, we admit that for a CTEG, multiple constraints could emerge for each n SISO TEG. We assume that there are Z capacity constraints applied to some places belonging to the Timed Event Graph of n index, and they are noted p_{s_n}. For each constrained place, we note respectively the following notations: M_{sn_0} is the initial marking, t_{zn} and $t_{z'n}$ are the delays, and they are corresponding to the input and output transitions of the place p_{s_n}. $x_{zn}x_{z'n}$ are the counter functions corresponding to these transitions. We denote by α_{s_n} the path connecting the control transition to the input transition of the constrained place. $\tau_{\alpha_{s_n}}$ is the sum of the delays through this path.

We denote the expression of marking constraints by the following inequality:

$$x_{zn}(t) \leq (b_n - M_{sn_0}).x_{z'n}(t) \text{ with } b_n \in \mathbb{N} \tag{19}$$

Theorem 2: A CTEG consists of N TEGs, whose evolution is given by Eq. (1), sharing a safe resource noted as $\tilde{p}_1$. It is object of multiple capacity constraints of the form (19), and it admits a global control law of the form:

$$u(t) = \bigoplus_{z=1}^{Z} u_z(t) \tag{20}$$

Having for the z constraint a control law denoted as $u_z(t) = F_z.x(t-1)$ such that $F_z = \left(b - M_{ij0} - (A^{\tau_{\alpha n}}B)_{j_n}\right).A^{\tau_{\alpha n}+1}(z_n{}', :)$ is the feedback that ensures each constraint of the form (19), if the conditions of the form (18) are verified for each index $z'n$ and zn of the constrained place p_{s_n}, with $\phi_s = \tau_{\alpha s_n} + 1$ for $s = 1, \ldots, S$.

Proof is provided in [9].

3.4 Closed-loop control with the (Min, +) algebra of timed Petri nets subject to mixed constraints

3.4.1 *Time constraints*

Temporal constraints are common restrictions for Discrete Event Systems, and their consideration during the control synthesis stage is a current problem that has motivated many researchers [1–3, 5, 10–18].

In this section, we are interested in the satisfaction of these constraints, more particularly in the critical time, especially in the case of tasks where the allocated time must be limited by a maximum bound $\tau_{ij}^{\max}$. Given a place denoted by p_{ij}, $\tau_{ij}^{\min}$ represents the minimal stay time of the token in this place. This is the time that is already taken in advance by the linear model. Since $\tau_{ij}^{\min} = \tau_{ij}$, then $\tau_{ij}^{\max}$ is the additional time condition that we look to satisfy.

Referring to [19], the expression for the time constraint is derived. It is represented by the following inequality:

$$x_i(t) \geq m_{ij0} x_j\left(t - \tau_{ij}^{\max}\right) \tag{21}$$

3.4.2 *Marking constraints*

Petri net feature marking assigns a nonnegative integer number to the places that refers to number of existing tokens. It also defines the dynamic behavior of a given system that varies according to certain transition firing rules. The tokens are interpreted as available resources such as material storage capacity in storage areas, memory capacity in network communication systems, and products to be processed and that are occupying an industrial production line. In the literature, several attempts have been carried out for the synthesis of a control satisfying the marking constraints of different kinds and for particular Petri nets [4, 8, 20–23]. However, each of these approaches suffers from a particular limitation.

We assume that the place illustrated in **Figure 3** is subject to a mixed constraint. Let m_{ij0} be its initial marking. $x_j(t)$ and $x_i(t)$ are, respectively, the counter functions associated with the transitions t_j and t_i until t time. m_{ij} is the marking available in the place p_{ij} until t time like $m_{ij} \in [e, b]$, which is equivalent to:

$$m_{ij}(t) = x_j(t) - x_i(t) + m_{ij0} \tag{22}$$

According to [8], the marking is bounded as follows $m_{ij}(t) \leq b$, and subsequently, we could get the following inequality:

$$x_j(t) - x_i(t) + m_{ij0} \leq b \tag{23}$$

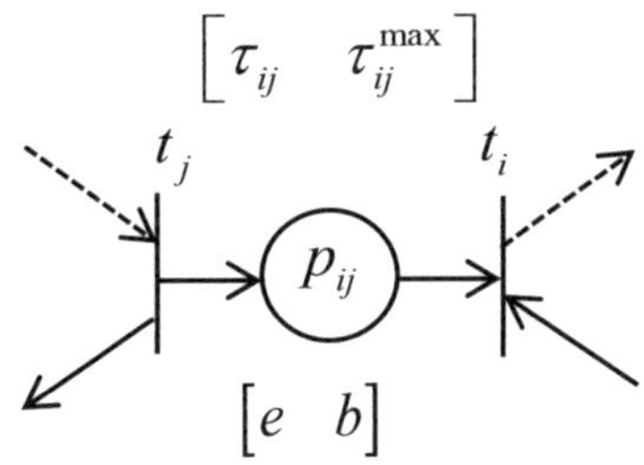

Figure 3.
Mixed constraints on the place p_{ij}.

and that could be transformed to its equivalent inequality (Min, +) as follows:

$$x_j(t) \leq (b - m_{ij0}).x_i(t) \tag{24}$$

4. Control formulation for TEGs with mixed constraints

In this section, we focus on the synthesis of appropriate control laws in the presence of mixed constraints on the places of the Timed Event Graphs. We start with the case of a single mixed constraint, and then, we generalize to the case of multiple constraints.

Considering the Eq. (9), we proceed by substituting the parameter τ by ϕ, and we the following equation:

$$x_i(t) = \left[\bigoplus_{r=1}^{N} (A^{\phi})_{ir} x_r(t-\phi)\right] \oplus \left[\bigoplus_{k=0}^{\phi-1} (A^k.B)_i.u(t-k)\right]; \phi \geq 1 \tag{25}$$

With $A \in \overline{\mathbb{R}}_{\min}^{n \times n}$ and $B \in \overline{\mathbb{R}}_{\min}^{n \times m}$, like n denotes the number of internal transitions, and m is the number of resource transitions.

The matrix given by $(A^{\phi})_{ir}$ denotes the rth components of the i row of the matrix A^{ϕ}.

To be more precise, we will opt for the notation $\phi = \phi_x$ in the case of the time constraint and the notation $\phi = \phi_y$ when it comes to the capacity constraint.

Let α be the path bounded by the source transition and the transition upstream of the mixed constrained place.

We define by τ_α and m_α successively the sum of all delays and the sum of markings existing along this path.

$x_j(t)$ represents the counter function of the transition upstream of the constrained place, and $u_x(t)$ represents the counter function of the source transition. As a result, we give the following inequality:

$$x_j(t) \leq m_\alpha.u_x(t - \tau_\alpha) \tag{26}$$

Remark 2: We admit that $u_x(t)$ represents the control that guarantees the time constraint case and $u_y(t)$ the control that guarantees the marking constraint.

4.1 Control synthesis of the TEG subject to a mixed constraint

TEGs are a major class within the Discrete Event Systems paradigm. This type of system seems to be more sensitive when the general process is exposed to mixed time and capacity constraints. We reserve this section for the introduction of the control synthesis in the case of existence of mixed constraints in the TEG.

Theorem 4.1: A TEG whose dynamics is given by the explicit equation of state representation, subject to time and capacity constraints on the place p_{ij} of the form (21) and (24) respectively. It admits a control law of the form $u(t)$:

$$u(t) = \min\left(F_x x_r(t-1), F_y x_r(t-1)\right) = \left(u_x(t) \oplus u_y(t)\right) = F_x x_r(t-1) \oplus F_y x_r(t-1) \tag{27}$$

With $F_x = \left[\bigoplus_{r=1}^{N} \left(A^{\phi_x}\right)_{ir} - m_{ij}.m_\alpha\right]$ such like $\phi_x = \tau_\alpha + \tau_{ij}^{\max} + 1$ and

$F_y = \left[\bigoplus_{r=1}^{N} A^{\phi_y}.\left(\left(b - m_{ij}\right) - m_\alpha\right)\right]$ such like $\phi_y = \tau_\alpha + 1$

This is also equivalent to:

$$u(t) = \min\left(\left[\bigoplus_{r=1}^{N} \left(A^{\phi_x}\right)_{ir} - m_{ij}.m_\alpha\right].x_r(t-1), \left[\bigoplus_{r=1}^{N} \left(A^{\phi_y}\right).\left(\left(b - m_{ij}\right) - m_\alpha\right)_{ir}\right].x_r(t-1)\right)$$
$$= \left[\bigoplus_{r=1}^{N} \left(A^{\phi_x}\right)_{ir} - m_{ij}.m_\alpha\right].x_r(t-1) \oplus \left[\bigoplus_{r=1}^{N} \left(A^{\phi_y}\right).\left(\left(b - m_{ij}\right) - m_\alpha\right)_{ir}\right].x_r(t-1)$$

If the following conditions are true:

$$F_y \geq 0; F_x \geq 0 \text{ et } m_\alpha \leq \left(b - m_{ij}\right).\left(A^k.B\right)_i \text{ Pour } k = 0 \text{ à } \tau_\alpha \tag{28}$$

Proof of this theorem is found [24].

4.2 Control synthesis of the GET subject to multiple mixed constraints

TEG might also be more sensitive when the same system is the subject of the presence of multiple mixed constraints. In this regard, since we have dealt with the case of control laws that satisfy a single mixed constraint, it seems important to deal with a more generalized problem that is multiple places subject to mixed constraints:

Let P_s be the places subject to multiple mixed constraints with $s = 1, \ldots, S$. These places are respectively delimited by the transitions $t_z, t_{z'}$, which are respectively associated to the counter functions x_z and $x_{z'}$. We denote by λ_z the cumulative delay from the control transition t_u to the input transition of the constrained place P_s.. For each place P_s, we define by m_z the initial marking. $\tau_z^{\max}$ is the maximum delay, and $b_z \in \mathbb{N}^*$ is the upper bound of the marking to be respected.

Consequently, the time constraints and the capacity constraints are given by:

$$x_{z'}(t) \leq m_z x_z\left(t - \tau_z^{\max}\right) \tag{29}$$
$$x_z(t) \leq (b - M_s) x_{z'n}(t) \tag{30}$$

Theorem 3: For a TEG whose evolution is described by the explicit equation of the state representation (1), it is subject to Z mixed constraints on the places of the form (29) and (30), admitting a general control law of the form:

$$U(t) = \bigoplus_{z=1}^{Z} u_z(t) \tag{31}$$

With:

$$u(t) = F.x(t-1) = \min(F_x.F_y)x_r(t-1) = (F_x \oplus F_y)x_r(t-1)$$

Like $F_x = \left[\bigoplus_{r=1}^{N} (A^{\phi_x})_{ir} - m_{ij}.m_\alpha\right]$ et $F_y = \left[\bigoplus_{r=1}^{N} (A^{\phi_y})_{ir}.((b - m_{ij}) - m_\alpha)\right]$ are the feedbacks that ensure the mixed constraints of the form (29) and (30), they are verified for each index z' and z of the constrained place p_s, with:

$$\phi_{x_s} = \tau_\alpha + \tau_{ij}^{\max} + 1 \text{ et } \phi_{y_s} = \tau_\alpha + 1 \text{ for each } s = 1, \ldots, S.$$

Proof of this theorem could be found in [24].

Considering that time and capacity constraints are two crucial factors for the successful and efficient conduction of these systems, we address the control problem for Conflicted Timed Event Graphs under mixed constraints. To this end, we build on the approaches introduced in [2, 8] to provide an adapted formulation of (Min, +) switching model policy.

4.3 Control of CTEGs subject to a mixed constraint with a single shared resource

Although many processes with nonpermissive standards, such as time and capacity criteria, are almost sensitive, they are found to be effective when these standards are met under proper system conduction.

At this level, we focus on the formulation of the control approach in the case of CTEGs exposed to a mixed time and capacity constraint. CTEGs are similar to competing users sharing a single resource among themselves. The solution of this problem is based on finding control laws with appropriate feedbacks that guarantee the mixed constraints applied on specific places of the CTEG.

Assuming that p_{ij} is the place subject to both a capacity and a time constraint, specifically located on the nth Timed Event Graph of the CTEG.

We first consider that this place is disposed to a temporal constraint presented by the interval:

$$\left[\tau_{ij}^{\min} \quad \tau_{ij}^{\max}\right].$$

$x_{j_n}(t)$ and $x_{i_n}(t)$ represent the counter functions of transitions t_j and t_i respectively, such that the firing number of these transitions is fixed until t time.

In the context of CTEG topic, the expression of the time constraint is expressed with reference to the n index of the TEG where it exists. It is given by the following linear inequality:

$$x_{i_n}(t) \geq m_{ij} x_{j_n}\left(t - \tau_{ij}^{\max}\right) \tag{32}$$

This same place may also be subject to marking constraint, which is given by the following inequality:

$$x_{j_n}(t) \leq (b - m_{ij})x_{i_n}(t) \tag{33}$$

Remark 3: In addition to the assumptions mentioned above, it is preferable to assume that each constrained place is accessible thanks to the index that determines its location in the Conflict Time Event Graph.

Remark 4: We assume that a control of the form $u(t) = F.x(t-1)$ is always a well-posed control law with delays and is preferred over a static control of the form $u(t) = F.x(t)$, such as $F \in \overline{\mathbb{R}}_{\min}^{m \times N}$. On the other hand, implicit control laws due to implicit loops for the controlled CTEG can lead to closed-loop blocking.

We provide in the following the equivalent equation to the explicit equation of state representation already given in (1).

$$x_x(t) = A^{\tau}.x(t-\tau) \oplus \left[\bigoplus_{k=0}^{\tau-1} A^k.B.u(t-k) \right], \text{with } \tau \geq 1 \tag{34}$$

By proceeding to the substitution $\tau = \phi$, like $\phi \geq 1$, one can achieve:

$$x_{i_n}(t) = \left[\bigoplus_{r=1}^{N'} (A^{\phi})_{ir}.x_r(t-\phi) \right] \oplus \left[\bigoplus_{k=0}^{\phi-1} (A^k.B)_i.u(t-k) \right] \tag{35}$$

Given the matrices $A \in \overline{\mathbb{R}}_{\min}^{N\prime \times N\prime}$ and $B \in \overline{\mathbb{R}}_{\min}^{N\prime \times m}$, such that N' represents the number of internal transitions of the Conflicted Timed Event Graph. The notation given by $(A^{\phi})_{ir}$ denotes the ith component of the matrix A^{ϕ} with r varying from 1 to N'.

For clarity purposes, $u(t-k)$ in Eq. (35) will be noted as $u_x(t-k)$ when we inspect for a control synthesis in the time constraint case and as $u_y(t-k)$ in the capacity constraint case, with α_n denoting the path bounded by the source transition and the upstream transition of the mixed constrained place.

For each path of the Conflict Timed Event Graphs in the chosen mode, we define respectively by τ_{α_n} and m_{α_n} the sum of all delays and the sum of the markings for each Timed Event Graph connecting the source transition to the upstream transition of the constrained place. Let t_{j_n} be its upstream transition of this place and x_{j_n} be its counter function. As a result, we obtain the following inequation:

$$x_{j_n}(t) \leq m_{\alpha_n}.u_x(t - \tau_{\alpha_n}) \tag{36}$$

In the case of a single shared resource, we begin by introducing Theorem 4 below in which we detail the control approach synthesis in the case of a single mixed constraint.

Theorem 4: A CTEG composed of N Timed Events Graph, whose evolution is given by the explicit equation of state representation, shares a safe resource denoted $\tilde{p}$. Let p_{ij} be the place subject to time and capacity constraints of the form (32) and (33) respectively. This CTEG admits a control law of the form $u_n(t)$:

$$u_n(t) = \min(F_x x_r(t-1), F_y x_r(t-1)) = (u_x(t) \oplus u_y(t)) = F_x x_r(t-1) \oplus F_y x_r(t-1) \tag{37}$$

Like:

$$F_x = \left[\bigoplus_{r=1}^{N} (A^{\phi_x})_{ir} - m_{ij}.m_{\alpha_n} \right];, \text{ we have } \phi_x = \tau_{\alpha_n} + \tau_{ij}^{\max} + 1$$

$$F_y = \left[\bigoplus_{r=1}^{N} A^{\phi_y}.\left(\left(b - m_{ij}\right) - m_{\alpha_n}\right)\right]; \phi_y = \tau_{\alpha_n} + 1$$

The above control is also equivalent to

$$u_n(t) = \min\left(\left[\bigoplus_{r=1}^{N} \left(A^{\phi_x}\right)_{ir} - m_{ij}.m_{\alpha_n}\right].x_r(t-1), \left[\bigoplus_{r=1}^{N} \left(A^{\phi_y}\right).\left(\left(b - m_{ij}\right) - m_{\alpha_n}\right)_{ir}\right]x_r(t-1)\right)$$

$$= \left[\bigoplus_{r=1}^{N} \left(A^{\phi_x}\right)_{ir} - m_{ij}.m_{\alpha_n}\right].x_r(t-1) \oplus \left[\bigoplus_{r=1}^{N} \left(A^{\phi_y}\right).\left(\left(b - m_{ij}\right) - m_{\alpha_n}\right)_{ir}\right]x_r(t-1)$$

The above control law exists if the following conditions are true:

$$F_x \geq 0, F_y \geq 0 \text{ and } m_{\alpha_n} \leq \left(b - m_{ij}\right).\left(A^k.B\right)_{i_n} \text{ for each } k = 0, \ldots, \tau_{\alpha_n} \tag{38}$$

Proof is found in [25].

4.4 Control of a CTEG with multiple mixed constraints

Switching systems represent a frequently encountered problem, especially in the context of real-time applications whose performance is very sensitive to the time factor and to the availability of the shared resource. In this case, the implementation of control laws that guarantee the respect of these conditions is required. It represents the motivation behind the study of this problem that deserves to be meticulously analyzed.

The following inequalities (39) and (40) define the expressions for the time and capacity constraints.

$$x_{z'}(t) \geq m_z x_z\left(t - \tau_z^{\max}\right) \tag{39}$$

$$x_{zn}(t) \leq (b_n - M_{sn}) x_{z'n}(t) \tag{40}$$

This section will be reserved to extend the previous case for the control of a CTEG in the case of the existence of multiple mixed constraints.

Theorem 5: For a CTEG constituted of N Timed Event Graphs, whose evolution is described by the explicit equation of state representation (1), shares a safe resource denoted $\tilde{p}$. Then, P_s are the places subject to z mixed constraints of the form (39) and (40). This CTEG admits a general control law of the form:

$$U(t) = \bigoplus_{z=1}^{Z} u_z(t) \tag{41}$$

Like $u_z(t) = F.x_r(t-1) = \min(F_x x_r(t-1), F_Y x_r(t-1)) = (F_x \oplus F_Y).x_r(t-1)$ such as $F_x = \left[\bigoplus_{r=1}^{N} \left(A^{\phi_x}\right)_{ir} - m_{ij}.m_{\alpha_n}\right]$ and $F_y = \left[\bigoplus_{r=1}^{N} \left(A^{\phi_y}\right)_{ir}.\left(\left(b - m_{ij}\right) - m_{\alpha_n}\right)\right]$ are the feedbacks that guarantee the mixed constraints of respectively the form (39) and (40). They are verified for each index z'_n and z_n of the constrained place p_{s_n}, having $\phi_{x_s} = \tau_{\alpha_n} + \tau_{ij}^{\max} + 1$ and $\phi_{y_s} = \tau_{\alpha_n} + 1$, for each $s = 1, .., S$. if the following conditions are true:

$$F_x \geq 0, F_y \geq 0 \text{ and } m_{\alpha_n} \leq \left(b - m_{ij}\right).\left(A^k.B\right)_{i_n} \text{ for each } k = 0, \ldots, \lambda_z \tag{42}$$

The demonstration of this theorem with an application case could be found in [25].

4.5 Generalization to the case of multiple shared resources

Multiple shared resources can be the object of multiple switches between the Timed Event Graphs constituting the CTEGs and performed according to the needs and specifications of the application. While multiple mixed constraints may also be involved, we move to a generalization for a control approach previously given in Section 4.4 that focuses on the existence of multiple users sharing multiple safe places while being a cause of the presence of multiple mixed constraints.

Remark 4: We note that at the modeling level, the mobilization of several shared resources is visible through the switching (Min, +) model within the sequence.

Theorem 6: For a CTEG consisting of N Timed Event Graphs, whose evolution is given by the explicit equation of state space representation (1), sharing multiple safe resources noted $\tilde{p}_l$ with $l = 1, \dots, L, \forall l \in \mathbb{N}$, it is subject to multiple mixed constraints given by the inequalities (39) and (40) respectively, admitting a global control law of the form:

$$U(t) = \bigoplus_{z=1}^{Z} u_z(t) \tag{43}$$

Having for z existing constraints on the places p_{s_n}, the control law in this case is noted $u_z(t) = \min(u_x(t), u_y(t))$, where $u_x(t) = \bigoplus_{r=1}^{N} A^{\phi_x}(z'_n, :).x_r(t-1) - m_{z'z}.m_{\alpha_n}$ and $u_y(t) = \bigoplus_{r=1}^{N} A^{\phi_y}(z'_n, :).x_r(t-1).((b_n - m_{iz'z}) - m_{\alpha_n}$ guarantee the existing constraints of the form (32) and (33), if and only if the conditions of the form (42) are verified for each index z'_n and z_n of the constrained places noted by p_{s_n}. having $\phi_{x_s} = \tau_{\alpha_n} + \tau_{ij}^{\max} + 1$ and $\phi_{y_s} = \tau_{\alpha_n} + 1$, for $s = 1, \dots, S$.

The proof is provided in [25] and an application case of a crossing railway sections is given.

5. Conclusion

Conflicted Timed Event Graphs turn out to be a rich class in the Discrete Event Systems paradigm with wide range of industrial applications. Based on their functionally critical nature where multiple users compete to win resource allocation, this type of systems seems to be more sensitive when the overall process is exposed to mixed time and capacity constraints. A wide investigation in this subject with multiple application cases referring to manufacturing applications and transport systems could be found in [26]. To this end, we detailed the switching model approach using the (Min, +) Dioid algebra. The conservative and mixed cases for resource allocation between graphs are considered beforehand. These contributions solve both sharing-resource-allocation problems, while the problem of the existence of capacity- and time-related marking constraints in places seems to be a source of complication for this type of system. Sufficient conditions are given for the existence of control laws.

Conflict of interest

The authors declare no conflict of interest.

Author details

Nesrine Ben Afia* and Hassani Messaoud
Electrical Department, National Engineering School of Monastir, Monastir, Tunisia

*Address all correspondence to: nesrine.benafia@enim.u-monastir.tn

References

[1] Cottenceau B, Hardouin L, Boimond J-L, Ferrier J.-L. Synthesis of Greatest Linear Feedback for Timed Event Graphs in Dioid. Automatic Control, IEEE Transactions on. 1999;**44**:1258-1262. DOI: 10.1109/9.769386

[2] Tebani K, Amari S, Kara R. Control of petri nets subject to strict temporal constraints using max-plus algebra. International Journal of Systems Science, Taylor & Francis Journals. 2018;**49**(6): 1332-1344. DOI: 10.1080/00207721.2018.1445311

[3] Van den Boom TJJ, De Schutter B. Modeling and control of switching max-plus-linear systems with random and deterministic switching. Discrete Event Dynamic Systems. 2012;**22**:293-332. DOI: 10.1007/s10626-011-0123-x

[4] Uzam M, Wonham W. A hybrid approach to supervisory control of discrete event systems coupling RW supervisors to petri nets. International Journal of Advanced Manufacturing Technology. 2006;**28**:747-760. DOI: 10.1007/s00170-004-2426-7

[5] Ben Afia N, Amari S, Messaoud H. A new formal method of control for Min-Plus linear systems subject to time constraints. In: International Conference on Control Automation and Diagnosis ICCAD. Grenoble, France: IEEE; 2019. pp. 1-6. DOI: 10.1109/ICCAD46983.2019.9037973

[6] Ben Afia N, Amari S. Synthetizing state feedback control Laws for discrete event systems with constrained paths using min-plus algebra. Studies in Informatics and Control. 2021;**30**:39-52. DOI: 10.24846/v30i1y202104

[7] Amari S, Demongodin I, Loiseau J-J, Martinez C. Max-plus control Design for Temporal Constraints Meeting in timed event graphs. IEEE Transactions on Automatic Control. 2012;**57**:462-467. DOI: 10.1109/TAC.2011.2164735

[8] Tebani K, Amari S, Kara R. State-feedback control for a class of timed petri nets subject to marking constraints. Asian Journal of Control. 2019;**21**(2): 934-951. DOI: 10.1002/asjc.1787

[9] Ben Afia N, Amari S, Messaoud H. Switching models and control of petri nets with shared resources under marking constraints. International Journal of Computer Integrated Manufacturing. 2021;**35**:113-128. DOI: 10.1080/0951192X.2021.1972463

[10] Amari S, Demongodin I, Loiseau JJ. Sizing, Cycle Time and Plant Control Using Dioid Algebra. In: Dolgui A, Soldek J, Zaikin O, editors. Supply Chain Optimisation. Applied Optimization. Vol. 94. Boston, MA: Springer; 2004. pp. 71-85. DOI:10.1007/0-387-23581-7_6

[11] Bonhomme P. Scheduling and control of real-time systems based on a token player approach. Discrete Event Dynamic Systems. 2013;**23**:197-209. DOI: 10.1007/s10626-012-0140-4

[12] Addad B, Amari S, Lesage J-J. Networked conflicting timed event graphs representation in (max,+) algebra. Discrete Event Dynamic Systems. 2012;**22**(4):429-449. DOI: 10.1007/s10626-012-0136-0

[13] Dideban A, Zareiee M, Hassane A. Controller synthesis with highly simplified linear constraints. Asian Journal of Control. 2013;**15**:80-94. DOI: 10.1002/asjc.528ff

[14] Dasarathy B. Timing constraints of real-time systems. Constructs for

expressing them, methods for validating them. IEEE Transactions on Software Engineering. 1985;**11**:80-86. DOI: 10.1109/TSE.1985.231845

[15] Miao L, Mao J, Cassandras CG. Optimal energy-efficient downlink transmission scheduling for real-time wireless networks. IEEE Transactions on Control of Network Systems. 2016;**4**(4): 692-706. DOI: 10.1109/TCNS.2016.2545099

[16] Mao J, Cassandras CG. On-line optimal control of a class of discrete event systems with real-time constraints. Discrete Event Dynamic Systems. 2010; **20**(2):187-213. DOI: 10.1109/TAC.2008.2009572

[17] Wang Y, De Schutter B, van den Boom TJJ, Ning B. Optimal trajectory planning for trains-a pseudospectral method and a mixed integer linear programming approach. Transportation Research Part C. 2013;**29**:97-114. DOI: 10.1016/j.trc.2013.01.007

[18] Ben Afia N, Amari S, Messaoud H. A new formal method of control for Min-Plus linear systems subject to time constraints. In: International Conference on Control Automation and Diagnosis ICCAD. Vol. 2019. Grenoble, France: IEEE; 2019. pp. 1-6. DOI: 10.1109/ICCAD46983.2019.9037973

[19] Amari S, Demongodin I, Loiseau JJ. Sizing, cycle time and plant control using Dioid algebra. In: Dolgui A, Soldek J, Zaikin O, editors. Supply Chain Optimisation. Applied Optimization. Vol. 94. Springer; 2005. pp. 71-85. DOI: 10.1007/0-387-23581-7_6

[20] Germano S, Laurent H, Jörg R. Optimal control of timed event graphs with resource sharing and output-reference update. Automatisierungstechnik. 2020;**68**:512-528. DOI: 10.1515/auto-2020-0051

[21] Ramadge P, Wonham W. Supervisory control of a class of discrete event processes. SIAM Journal on Control and Optimization. 1987;**25**(1): 206-230

[22] Tebani K, Amari S. Min-plus realizable control design for partially observable timed event graphs under marking constraints. European Journal of Control. 2021;**57**:33-40. DOI: 10.1016/j.ejcon.2020.12.002

[23] Yang B, Hu H. Implementation of generalized mutual exclusion constraints using critical places and marking estimation. In: IEEE Transactions on Systems, Man, and Cybernetics Systems. 2019;**51**:2168-2232. DOI:10.1109/TSMC.2019.2944886

[24] Ben Afia N, Amari S, Messaoud H. Contribution on the control of mixed constrained discrete event Systems for a Flexible Workshop Application. Journal of Advances in Information Technology. 2022;**13**(1):9-14. DOI: 10.12720/jait.13.1.9-14

[25] Ben Afia N, Amari S, Messaoud H. Control of conflicted timed event graphs subject to mixed constraints using (Min, +) semiring-Application to a railway network. IFAC Journal of Systems and Control. Nov 21-2022

[26] Nesrine Ben Afia. Modeling and Control of Conflicted Timed Events Graphs under Constraints Using Dioid Algebra (Min, +) [thesis]. Tunisia: National Engineering School of Monasir: Monastir University; 2022

Chapter 6

Review of Model Predictive Control and Other Methodologies for Lateral Control of an Autonomous Vehicle

Francis Assadian and George Martin

Abstract

This chapter summarizes controller methodologies for lateral control of a standard passenger vehicle, with front-wheel steering and fixed rear wheels. Comparisons are made to elucidate benefits and drawbacks of the application of a model predictive controller. Select widely used system models are described, and a general problem formulation for the model predictive controller is given. Two publicly available examples of this application are reviewed and compared, and the results are discussed.

Keywords: control, model predictive control, vehicle dynamcis, autonomous vehicles, lateral control

1. Introduction

Automatic navigation and control of terrestrial vehicles are a rapidly growing and developing field, with a wide range of applications. From meal-serving robots in restaurants to GPS-localized tractors that can till fields without human interaction, the potential for this domain of research and application in industry is only recently being expanded. Recent advances in computational, sensory, and battery technologies have enabled the automation of perhaps the most popular domain, automotive systems. The automotive industry has a long history of utilizing feedback control systems, in functions such as cruise control for longitudinal control and lane-keeping assistance for lateral control. In fact, control of an autonomous vehicle has historically been decomposed into these two categories – longitudinal and lateral control. However, in order to achieve higher levels of autonomy, it is vital to develop a control framework that generates a high-quality coordinated response between both vehicle states. In this chapter, a comparison between the various control schema will be illustrated to weigh the pros and cons of a model predictive controller in this application. For the use of an MPC system in simulation, it will be necessary to compare several models, MPC formulations, and optimization methods. At the end of this chapter, one example of such a controller will be presented for both a linear and nonlinear case to illustrate the performance and capabilities of such a controller, as well as the advantages and disadvantages, which come with its implementation.

2. Controller review

In the application of automotive controllers, virtually all inputs can be described as a combination of a yaw steering input produced by the steering wheel, and a velocity target input produced by the accelerator pedal. There do exist some controllers that can utilize torque vectoring or incorporate additional actuators to the system to add an additional degree of freedom, but these will not be addressed in this chapter as they require equipment not found on the vast majority of passenger vehicles. Controllers can be categorized in several ways, and for simplicity's sake will be compared within three major categories: geometric, frequency domain, and state-space formulated controllers. Controllers in the latter two categories have some overlap, in that linear controllers can almost always be described in both a state-space and frequency domain representation; the distinction in this section is made to provide a *concise* description of the different methodologies. This chapter will focus on the lateral control of a standard passenger vehicle but will include consideration for longitudinal control where relevant and applicable. Controllers are usually given a reference trajectory, which is the desired path for the vehicle to take and must extract the useful information from this path and the vehicle's relative pose to generate the controller outputs.

2.1 Geometric controllers

Geometric controllers simply use the geometric and kinematic relations of the vehicle to generate an output. In lateral control, the most popular formulations are the pure-pursuit and Stanley controllers.

2.2 Pure-pursuit

Perhaps the most simple controller that sees real-world use, a pure pursuit controller uses some look-ahead distance to generate a steering output. In the most simple application, a fixed distance is selected, and a point on the target trajectory path at this distance from the car is used to determine the steering output. The output steering is directly proportional to the yaw angle of this geometric link. Additional complexity can be added by having this look-ahead distance vary as vehicle speed changes. This works well in simple, low-curvature environments but has no claims of optimality or robustness. Furthermore, as the vehicle approaches a path, such as a straight line or slight curve, it will asymptotically approach the line even in ideal circumstances. Additional logic needs to be included to improve target following and account for situations where the car is further away from the path than the look-ahead distance.

2.3 Stanley

The Stanley controller is a more advanced geometric controller, which uses a function that is shown to be stable within a vehicle's operating range. The most basic form of the Stanley function uses the current vehicle yaw $\psi(t)$, lateral error $e(t)$, and velocity $v(t)$ along with a gain k to generate a steering output $\delta(t)$ [1].

$$\delta(t) = \psi(t) + arctan\,\frac{ke(t)}{v(t)} \tag{1}$$

This controller is able to operate within a wider operating range than a pure pursuit controller, similar computational requirement, and ease of implementation. Additionally, only one gain is required to be tuned, although the formulation can be modified with other gains and considerations in practice. However, this controller is highly dependent on a well-tuned yaw sensor as beyond a certain threshold noise can significantly impede performance. Furthermore, this controller suffers in its ability to provide adequate robustness.

2.4 Frequency domain

Controllers developed using frequency domain analysis have a rich history of use due to their ability to produce optimal and robust linear controllers, as well as the ease of implementation in computationally constrained environments.

2.5 PID

$$u(t) = K_p e(t) + K_i \int e(\tau) d\tau + K_d \frac{de(t)}{dt} \tag{2}$$

PID controllers are the most popular general controller, due to their performance on a wide variety of systems, simple formulation as a second-order low pass filter, and procedural tuning methodology. The proportional, integral, and derivative gains are able to correct the input based on the current error, sum of previous errors, and rate of change of the error. For many systems, this is sufficient to produce the desired performance. This method can be tuned directly without a system model, or the controller gains could be derived utilizing an appropriate system model.

2.6 Youla-Kucera or Q parametrization

Youla or Q parametrization is a technique that uses a plant model, linearized if necessary and converted to the frequency domain, to generate a stabilizing controller. The additional knowledge of the system can allow the Youla controller to accommodate more complex dynamics, and even allow the designer to choose between under-damped, over-damped, and critically-damped response behavior depending on their desired application [2]. Additionally, this method can be applied to multiple-input--multiple-output (MIMO) systems and, with some constraints in the formulation, generates a controller, which accommodates complex behavior between the inputs and outputs of a system.

2.7 H_2 and H_∞

The H_2 and H_∞ controllers use the concept of norm minimization in the frequency domain to develop a controller that guarantees performance of a linear system [3]. H_2 controllers minimize a quadratic cost function or 2-norm of a system, which can be shown to guarantee performance even with Gaussian noise interference to the system. H_∞ controllers use a multiplicative gain to shape the closed-loop transfer function and serve to mitigate worst-case scenarios and guarantee stability.

2.8 State space

State-space controller formulations use a time-based approach, as opposed to frequency domain. The differential equations governing the system are more intuitive for an engineer to comprehend the modeled behavior of the system, and for linear systems, this model can be described in matrix form, which allows for efficient controller computation. The drawback to this approach is that the frequency response is not directly considered, and the controller performance depends on the quality of the model in question.

2.9 LQR and LQG

LQR or **l**inear **q**uadratic **r**egulator design uses a state-space formulation, with a quadratic cost function, to produce a set of fixed feedback gains *K*, which produce an optimal response. This requires the system to be linear time-invariant, but given these conditions can produce linear optimal controllers for both SISO and MIMO systems. An LQG or **l**inear **q**uadratic **G**aussian controller expands upon this formulation and produces an optimal controller, which considers Gaussian noise of a known or estimated distribution by the addition of a Kalman filter. However, in order for a system without full-state feedback to be guaranteed robustly stable, loop transfer recovery methods must be employed.

2.10 Control-Lyapunov function

A control Lyapunov function is a function that can prove mathematically that a system is asymptotically stabilizable, and can be used to generate a controller, which is guaranteed to perform within the state-space. These functions do not have specific methods to find and can prove difficult to determine in complex systems. However, Lyapunov functions for linear systems can always be written as quadratic functions, and there is a significant overlap with LQR controllers in this sense. Lyapunov functions add the ability to generate a provable stabilizing controller for nonlinear systems in all state-space if they can satisfy the criterion, which makes this method attractive in the domain of nonlinear control [4]. Unfortunately, Lyapunov functions are very difficult to find, which satisfy the criterion, and it is possible that for a given system this function may not exist. For this reason, this style of controller has seen limited use.

2.11 MPC

Model predictive controllers have recently experienced a renaissance in research due to the rapid development of computational technology to the point that real-time MPC has become feasible in a wide set of applications, one such use being automotive control. MPC controllers discretize the state-space and forward project the systems state up to a time horizon and then optimize a fixed response within each time step to produce an optimal output. This method is, therefore, able to forward simulate and incorporate physical actuator limits. In the case of these nonlinear systems, convergence becomes more challenging to guarantee. Convergence will be discussed in a later section. The primary drawback of MPC controllers is the massive computational cost, which has become less difficult to overcome with each passing year. Additionally, a significant amount of effort has been and is being done to create highly computationally efficient controllers that are designed for their respective applications.

2.12 Why MPC?

Due to the wide variety of controllers applicable to the problem of vehicle control, one must weigh the benefits and drawbacks of any controller to be implemented against all others to determine which best fits the vehicle's intended function. For this reason, it is useful to describe the specific utility provided by MPC and justify the cost of including more powerful computational equipment onboard a vehicle. Model predictive control methods are able to approximate an optimal solution to the control problem, which incorporates realistic actuator and state constraints. Additionally, these controllers are able to forward-project, which is highly useful in automotive applications as the controller can account for path constraints and obstacles.

Model predictive controllers, while difficult to *prove* as robust, are nonetheless able to generate optimal outputs that are updated in response to the actual dynamics of the system. They also introduce a critical consideration in optimization that controller actuators are constrained not only within minimum and maximum bounds but within transition speed limits as well. Since controllers use a constrained optimization technique at each time step, the capabilities of the actuators can always be included in the problem and produce outputs that are both optimal and feasible.

3. Modeling

Perhaps the most popular model format used in the control of cars and car-like robots is the bicycle model. The widespread use of this modeling style can be attributed to its ability to accurately describe the motion of a vehicle across a variety of maneuvers, using a minimal set of states. The general assumption of this model is that there is no weight transfer or roll in the vehicle and that both front and rear sets of wheels can be effectively reduced to a single wheel in the front and rear of the vehicle for control purposes. In this section, controllers of increasing complexity will be described, which are based on these original assumptions.

3.1 Kinematic bicycle model

In a kinematic bicycle model, dynamic interactions between the tires and the road are ignored, and perfect traction is assumed. For a standard passenger vehicle, the vehicle is assumed to have a front wheel that is free to rotate, and a fixed rear wheel. Essentially, the vehicle is modeled as a two-dimensional bicycle, which is graphically represented in **Figure 1**.

$$\begin{aligned}
\dot{x} &= vcos(\psi + \beta) \\
\dot{y} &= vsin(\psi + \beta) \\
\dot{\psi} &= \frac{v}{l_r}\, sin\,(\beta) \\
\dot{v} &= a \\
\dot{\beta} &= arctan\left(\frac{l_r}{l_f + l_r}\, tan\,(\delta_f)\right)
\end{aligned} \tag{3}$$

This model is inherently nonlinear due to the rotation of the front tire as an input and its corresponding impact on the system as a whole. As such, it is sometimes useful

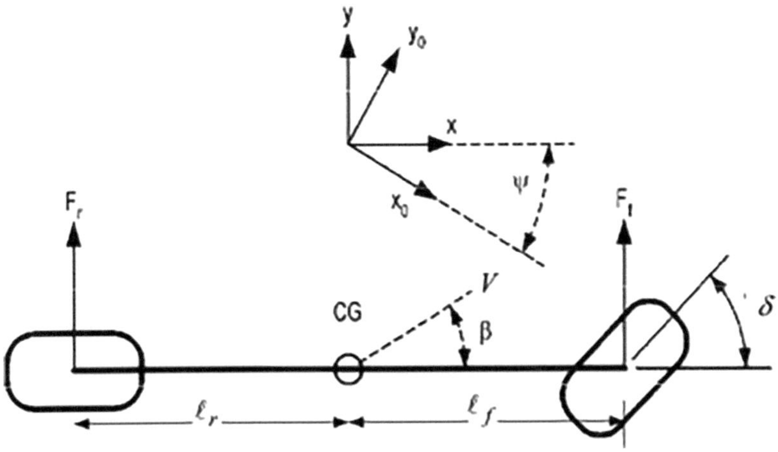

Figure 1.
Diagram of bicycle model [5].

to linearize this system about a zero steering angle in order to develop the linearized kinematic bicycle model. This model describes the kinematic motion of a front-steered vehicle well but makes one assumption that greatly limits its performance. The kinematic model's assumptions imply that the vehicle's front and rear wheels will drive forward in exactly the direction they are aimed, which is not true, especially at larger steering angles. In reality, tires flex and the contact patch where the tire meets the road can distort, leading to behavior that does not fit this model. At higher lateral accelerations, tire dynamics become one of the most significant parameters in lateral control.

3.2 Dynamic bicycle model

The dynamic model is an expanded form of the kinematic model, in that tire forces are now included. Tire behavior is nonlinear and does saturate beyond a certain threshold. The most famous model used in both academia and industry is the Pacejka Magic Tire formula, which was empirically derived from a series of experiments designed to capture pure cornering motion [6]. Though there is not a direct link to known physical phenomena within the tire, the Magic Formula accurately captures tire behavior over a wide range of conditions.

$$F_y(\alpha_y) = D_y \, sin\left(C_y \, arctan\left(B_y * \alpha_y - E_y\left(B_y\alpha_y - \, arctan\left(B_y\alpha_y\right)\right)\right)\right) + S_{V_y} \text{where} \alpha_y = \alpha + S_{H_y.} \quad (4)$$

In this model, coefficients A, B, C, D, and E are curve fit to a set of experimental observations, while the S parameters are linear shifts. These parameters remain valid for a vehicle so long as tire load F_z and camber remain constant. For a bicycle model, this does not add any additional assumptions. This model can be directly incorporated as the lateral tire force, or if desired linearized about the origin to produce the more simple relation.

$$F_y = C_y\alpha_y \quad (5)$$

This assumption simplifies the model but loses the ability to model tire saturation, and therefore, only performs adequately within a near-linear range. Below a lateral acceleration of 4–5 m/s^2, this model is sufficient for most applications. When incorporated into the linear bicycle model, the state space notation for lateral motion of a bicycle model becomes

$$\begin{bmatrix} \dot{v} \\ \dot{r} \end{bmatrix} = \begin{bmatrix} -\dfrac{C_r + C_f}{mu} & -u - \dfrac{C_f a_1 - C_r a_2}{mu} \\ \dfrac{-C_f a_1 + c_r a_2}{u I_z} & \dfrac{-C_f a_1^2 - C_r a_2^2}{I_z u} \end{bmatrix} \begin{bmatrix} v \\ r \end{bmatrix} + \begin{bmatrix} \dfrac{C_1}{m} \\ \dfrac{C_f a_1}{I_z} \end{bmatrix} \delta_f \tag{6}$$

$$y_e = \begin{bmatrix} 0 & 1 \end{bmatrix} \begin{bmatrix} v \\ r \end{bmatrix}.$$

3.3 Additional higher complexity models

Additional higher-order models of note include the planar and double bicycle models, which model each tire with a fixed track width between two linked bicycle models, and the full model, which can consider roll and weight transfer. There are a limitless number of considerations that can be made, but each additional required state exponentially increases the computation time required for an MPC controller to generate the optimal solution, and as of yet are not feasible without a significant investment into the computational platform. The following examples will use more simple models for ease of explanation as well as viability for use without heavy resource commitments.

4. Model predictive control of an autonomous vehicle

In recent years, model predictive control has become of great interest in autonomous vehicle engineering, as it is able to formulate an optimal control law given realistic vehicle constraints, and adjust in real-time to a change in target trajectory as well as vehicle dynamic conditions. It is adaptable to a wide variety of applications and provides a great deal of flexibility to the control designer. However, the greatest limiting factor is the computation time required to achieve an optimal solution. In practical applications, significant time must be given to ensure that the controller will achieve good performance and provide commands quickly enough to function in a real-world environment. A rule of thumb in this application is the outermost loop of the controller should iterate at roughly 10 Hz frequency, which has been used several times in academic studies [7, 8]. Currently, computing technology has advanced to the degree that both linear and nonlinear MPC controllers are beginning to see use. In this section, the MPC formulation of a vehicle controller will be elucidated, and considerations will be noted with regard to realistic functional development (**Figure 2**).

4.1 MPC general problem—Structure

The general formulation of a model predictive controller in the sense of automotive control is to solve a constrained optimization problem to generate optimal steering and velocity commands, which can reliably follow a reference trajectory. This optimization is performed by minimizing a cost function, which enables the designer to tune the system's response intuitively by weighing the significance of different states, as well as state transition speeds and a variety of other factors. The general formulation of a continuous-time optimal control problem is as follows:

$$
\begin{aligned}
&\underset{x,u}{minimize} \int l_c(x(t), u(t))dt + V_f x(T) \\
&\text{subject to :} \quad x(0) = x_0 \\
&\qquad \dot{x}(t) = f(x, u) \quad t \in [0, T] \\
&\qquad (x(t), u(t)) \in \mathbb{Z} \quad t \in [0, T] \\
&\qquad x(T) \in \mathbb{X}_f
\end{aligned} \tag{7}
$$

In this formulation, l_c is a cost function of the states and inputs, while V_f is a terminal constraint associated with the final state in the time horizon. This general formulation assumes that the states are defined only as a function of states and inputs, as well as that the states and inputs are bounded in some finite space, as is the terminal state.

This problem in its current state is unwieldy, as in continuous time, there are an infinite set of states x and u in the considered time range to be optimized. There are two main classes of methods, known as direct and indirect methods. While both methods have merits in their own right, direct methods tend to produce a solution more quickly (by sacrificing the guarantee of optimality). The primary difference is that indirect methods will optimize some approximate results in continuous time, then convert this solution to a set of discrete commands, while direct methods discretize the system first before solving the optimization problem. Since speed is critical, this chapter will focus primarily on direct methods. To begin, the system must be discretized to become solvable within a meaningful length of time. Discretization methods vary slightly depending on the choice of optimization variables, but there are a few requirements for all discrete systems used in MPC.

Time step—The most basic component of a discrete system, the continuous system is subdivided by an equal time step. Smaller time steps imply a larger computational cost but also a corresponding increase in resolution of the control output.

Time horizon—MPC employs a receding horizon, that is to say that each iteration of the outer loop minimizes within a fixed time horizon. In simple terms, if the receding horizon is 10 seconds, the algorithm will minimize the cost function for the set of all time steps within the next 10 seconds from the current time. The time horizon should at minimum be selected to be sufficiently long that the controller can project a trajectory that converges to the path where possible, as a shorter horizon time can result in a non-stabilizing controller.

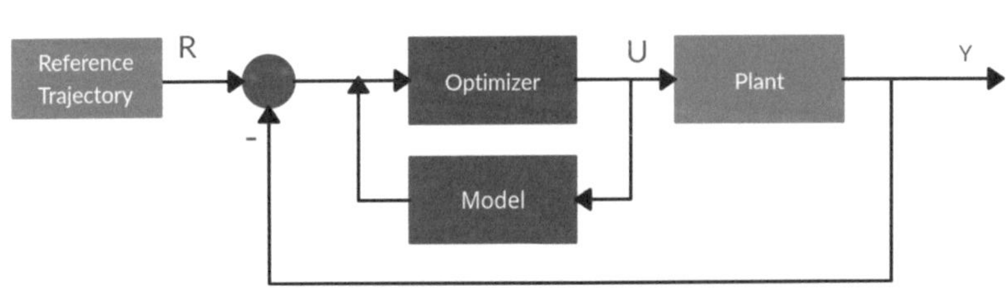

Figure 2.
MPC loop structure.

Input assumption—For practical MPC controllers, the input u is considered to be piecewise constant within each time step, sometimes referred to as the step input condition. In some cases, this can be relaxed but for the purposes of this chapter, other conditions will not be considered as they require a different approach to optimization.

4.2 Formulation and transcription

For reasons that will be clear to those with experience using cost functions, and that will be explained later in the section, it is practical and many times ideal to use a quadratic cost function similar to an LQR formulation. In the example below, there is a quadratic cost for the states, inputs, derivative of inputs, and final value.

$$J = x^T Qx + u^T Ru + \dot{u}^T R_\Delta \dot{u} + \Phi(x_f) \tag{8}$$

This cost function is a specific implementation of the general formulation described earlier, in that Q describes a quadratic state cost, R describes a quadratic input cost, R_Δ describes a quadratic cost to a change in input, and ϕ describes a scalar terminal cost. In comparison to the general form, J replaces l_c and V_f in the optimization problem.

Using the assumption that inputs within a time step are piecewise constant, with such a cost function, it is possible to convert the boundary value problem to an initial value problem; that is to say that an optimizer needs to solve for a constant input that minimizes the cost function across its specific time step. This is referred to as a shooting method and has the greatest potential for real-time applications.

4.3 Direct shooting methods

Direct shooting methods are the simplest methods that can be applied to produce a performant output, and exploit the formulation described in previous sections. Since the boundary problem has been converted to an initial value problem, an optimizer can, within each time step, guess and check the input value and observe the output, moving closer to the optimal value with each iteration. The solving method by which this is guaranteed to approach an optimal solution will be described in the next section, as well as the constraints imposed on the initial value optimization method. Once an optimal solution is found, its final state values are used to generate the next initial value problem in the set. This process is repeated until the time horizon is reached.

Direct multiple shooting is an extension of direct single shooting [9]. Instead of solving the first time step, then continuing forward in time to reach a solution, direct multiple shooting methods solve each time step in parallel, allowing a discontinuity between the time steps to appear—this is called a defect. The defect is then included in the optimization problem as a term to be minimized, and thus after iteration becomes negligible. If an optimal solution is reached, the absolute sum of defects should approach zero. Now that the transcription method and optimization problem have been formulated, the only remaining component is how the optimization problem is actually solved.

5. Solving the optimization problem

Direct methods require the use of an iterative solver, which approaches the final result in finite time. In this section, a few selected solver methodologies will be described and compared, starting with a basic solver to describe their fundamental behavior, before moving on to more complex methods. An important clarification is to be made with regard to optimization and its degree of difficulty. While distinction is generally made between linear and nonlinear MPC, the more significant factor determining the difficulty of optimization is whether the system is convex or non-convex. This is the primary reason that quadratic cost functions are preferred, which was referenced earlier in the chapter. Quadratic cost functions maintain convexity in the solution space. Linear MPC with a quadratic cost function will always be a convex optimization problem, while nonlinear MPC *may* or *may not* be convex. Convexity in this context assumes that all points between two points in the set are also points in the set. This enables the use of interior point methods to generate a solution, which comprises the driving functions behind most popular optimization solvers used in MPC. Additionally, a convex set provides a guarantee that the solution will be globally optimal [10]. Optimization for non-convex problems introduces the risk that the solution will only be locally optimal, but there are techniques to mitigate this issue without making the computation time unrealistic for use in real-time applications like autonomous vehicle control.

One popular method for optimization is the Newton–Raphson method. This gradient descent method is a simple implementation that provides a good basis for understanding other optimization techniques. The formula for selecting the next iterative point is shown below:

$$x_{k+1} = x_k - \left(\nabla^2 f(x_k)\right)^{-1} \nabla f(x_k). \tag{9}$$

This method uses the Hessian of a function to approximate the global minima, as this in a sense minimizes a quadratic cost function. One benefit of this method is that it works very well for linear systems without a large number of states, with low relative computational cost. However, significant nonlinear behavior or a large number of states may require more advanced methods. Optimization is a large field with a wide breadth of literature so will not be covered in depth here, but the Newton–Raphson method provides a good intuitive sense of the general idea of how optimization is performed in a model predictive controller.

6. Simulations

In order to provide a more concrete and understandable model, two MPC simulations were selected to present their performance. There will not be a one-to-one comparison as each controller has a different application, but a curious reader will be able to use the available GitHub repositories to investigate beyond the phenomena and commentary provided here.

6.1 Linear MPC

A simple example of linear MPC using a kinematic bicycle model is available at the PythonRobotics GitHub repository [11]. The available model predictive speed and steering control utilizes a cubic spline planner to produce a set of ordered waypoints, which include an x, y, and yaw pose as well as velocity target. The simulation uses the direct single-shooting method based on the same cost function described earlier in this

Figure 3.
PythonRobotics linear kinematic model (5 m/s and 10 m/s, respectively).

chapter. The map used in this example is derived from the nonlinear example presented next in this section (**Figure 3**).

In this example, the MPC controller was modified to have a fixed longitudinal velocity, in order to accentuate the lateral behavior of the vehicle. In this linear example, which was simulated with a kinematic system, the vehicle is able to track the target path, however, performance is hindered at higher speeds. The most significant overshoot is observed in the lower left section of the track, where the linear model is unable to generate an ideal trajectory for a high curvature. Additionally, this model is able to effectively control for cross-track error and yaw rate but is limited in practical scope as it cannot make use of additional road information.

A much more advanced example from ETH Zurich presents a nonlinear model predictive control scheme, which incorporates a set of realistic constraints to the vehicle and path, and simulates a higher-order model, a dynamic bicycle model with Pacejka tire coefficients [12]. Again, this simulation was restricted to only allow for lateral control and was operated at a fixed speed of 5 and 10 meters per second. However, this example has a more developed cost function, and as such select weights are compared to differentiate between a forward-progression optimization and path-tracking optimization.

$$
\begin{aligned}
J &= e_c^T q_c e_c + e_l^T q_l e_l + \Delta u^T R_\Delta \Delta u \\
\text{Subject to :} \quad & x(0) = x_0 \\
& \dot{x}(t) = f(x,u) \\
e_l(x) &= \sin\left(\phi_{ref}\right)\left(X - X_{ref}\right) - \cos\left(\phi_{ref}\right)\left(Y - Y_{ref}\right) \\
e_c(x) &= -\cos\left(\phi_{ref}\right)\left(X - X_{ref}\right) - \sin\left(\phi_{ref}\right)\left(Y - Y_{ref}\right) \\
& x \in \mathbb{X}_{track}
\end{aligned}
\tag{10}
$$

This formulation is still quadratic but uses a lag cost term q_c as well as state, input, and terminal cost. The state costs are also bounded within a track, which constrains the optimization such that the vehicle cannot predict a path that leaves the track (**Figure 4**).

Figure 4.
Modified MPCC controller simulation at select speeds and optimization weights.

Tracking optimization is performed by increasing relative weight to both of the error states e_l and e_c, while progression optimization is made by increasing the relative weight of the lag cost parameter e_l included in this cost function. This presents a functional means to tailor the cost function to the specific vehicle application. For an average passenger vehicle, the performance of the tracking optimized cost functions would serve as more useful while a race car would benefit from the increased emphasis on forward progress over strict path tracking. Additionally, the nonlinear model is able to benefit from the characteristics of the tire to achieve very precise motion around the tight lower left bend, while avoiding collision with the side of the road due to the constraints placed on the controller. The PythonRobotics method, using single shooting without an optimization library, has a controller frequency of about 10 Hz, while the MPCC (model predictive contouring control) method was able to provide commands at nearly 200 Hz on the same platform. This is due to the implementation of multiple shooting as well as the inclusion of a performance-based optimization library.

7. Conclusions

The comparison presented in this chapter illustrates the importance of implementation when designing an MPC controller. Though the PythonRobotics model is simpler and possesses fewer states, it runs significantly slower than the MPCC controller. This is primarily due to two key factors—coding language and optimization style.

Direct multiple shooting converges faster than single shooting due to the increased dimensionality of slack parameters between each time step. Furthermore, Python is inherently a slower language than C++ because is it not a compiled language. In the application of autonomous vehicles, where fast control response is essential, it is of the utmost importance to formulate the problem and design using languages that reduce the computational time wherever possible.

The benefits of model predictive control are additionally shown in the simulated examples above, especially the MPCC controller, which utilizes bounded optimization to navigate the track near the edge of the drivable terrain. Without an intelligent path planner, it is difficult to design a controller through any other means that is able to use information about the drivable space and optimize the control response to take advantage of it. While this effect is achievable with more complex path planning and replanning, model predictive control is able to provide a collision-free output whose performance is dependent upon the model itself rather than an external planning algorithm. More advanced autonomous systems are additionally able to forward-project moving objects, and a model predictive controller can take these projections and generate an online output that navigates around such objects.

In existing real-world applications, longitudinal control is additionally incorporated in order to have the vehicle achieve a form of full-optimized piloting. This provides a significant advantage in some cases as many other controllers require either a decoupling of the longitudinal and lateral control or some modification to the control law to achieve good performance. The increased computational cost of this controller is quickly becoming a smaller and smaller price to pay for this performance, and the nonlinear model predictive control of a passenger vehicle is becoming increasingly feasible to implement. Collision avoidance, behavioral prediction models, and stochastic prediction are just a few of the many possible improvements that can be incorporated to achieve control with a well-defined set of safety considerations. This will ultimately provide an extra layer of safety and performance in the computational stack of a self-driving vehicle, beyond just following a dictated path.

Code GitHub Repositories

https://github.com/AtsushiSakai/PythonRobotics
https://github.com/alexliniger/MPCC

Author details

Francis Assadian* and George Martin
University of California, Davis, USA

*Address all correspondence to: fassadian@ucdavis.edu

References

[1] Hoffmann GM et al. Autonomous automobile trajectory tracking for off-road driving: Controller design, experimental validation and racing. In: American Control Conference, New York, NY. 2007. DOI: 10.1109/acc.2007.4282788

[2] Assadian F, Mallon KR. Robust Control: Youla Parameterization Approach. Hoboken, New Jersey: John Wiley & Sons Ltd; 2022

[3] Rafaila RC, Livint G. H-Infinity Control of Automatic Vehicle Steering. In: 2016 International Conference and Exposition on Electrical and Power Engineering (EPE), Iaşi, Romania. 2016. DOI: 10.1109/icepe.2016.7781297

[4] Khalil HK. Nonlinear Systems. 3rd Edition. Upper Saddle River: Prentice Hall; 2002

[5] Rajamani R. Vehicle Dynamics and Control. 2nd Edition. Berlin: Springer; 2012. DOI: 10.1007/978-1-4614-1433-9

[6] Pacejka HB, Bert CW. Tyre models for vehicle dynamics analysis. Journal of Applied Mechanics. 1995;**62**(1):268-268. DOI: 10.1115/1.2895925

[7] Zeng J et al. Safety-Critical Model Predictive Control with Discrete-Time Control Barrier Function. In: American Control Conference (ACC), New Orleans, LA. 2021. DOI: 10.23919/acc50511.2021.9483029

[8] A Predictive Path-Following Controller for Multi-Steered Articulated … https://arxiv.org/pdf/1912.06259v4

[9] Bock HG, Plitt KJ. A multiple shooting algorithm for direct solution of optimal control problems *. IFAC Proceedings Volumes. 1984;**17**(2):1603-1608. DOI: 10.1016/s1474-6670(17)61205-9

[10] Schwenzer M et al. Review on model predictive control: An engineering perspective. The International Journal of Advanced Manufacturing Technology. 2021;**117**(5–6):1327-1349. DOI: 10.1007/s00170-021-07682-3

[11] Sakai A et al. PythonRobotics: A Python Code Collection of Robotics Algorithms

[12] Liniger A et al. Optimization-Based Autonomous Racing of 1:43 Scale RC Cars. Optimal Control Applications and Methods. 2014;**36**(5):628-647. DOI: 10.1002/oca.2123

Chapter 7

Implementation of MPC Strategy in Reactive Separation Techniques and Its Benefits: A Demonstration with Natural Gas Sweetening Process

K. Karthigaiselvan and Rames Chandra Panda

Abstract

Model Predictive Control (MPC) is a widely used method that has numerous applications in process industries. In the MPC group of controllers, a clear process model is used directly for controlling, predicting future plant behavior, and calculating corrective control action required to maintain the output at the desired set point value. Most chemical processes exhibit inherent nonlinearities due to interactions among processes, disturbance, and set-point changes. MPC variants based on nonlinear process models have been a proven stiff control technique for process control along with improved handling of constraints, abnormal dynamics, and time delays. In addition to that, MPC is better in handling the nonlinearity and time-varying characteristics during run time by modifying model. The control of multi-input and multi-output reactive-separation process is difficult due to nonlinearity associated with the process and interactions of vapor-liquid equilibrium with chemical reactions. In order to obtain optimal performance, energy conservation, and cost-effectiveness in reactive-separation process, the application of optimal control technique is inevitable. This chapter addresses application of MPC and its benefits in reactive separation techniques, particularly in natural gas sweetening process. The recent application in MPC and its proven results for the above-mentioned reactive separation processes are discussed here.

Keywords: reactive absorption, conventional controller, MPC, performance specifications, natural gas sweetening process

1. Introduction

Process Integration (PI) deals with the improvement of innovative equipment and techniques that fetches profound advancements in the field of chemical process and manufacturing industries. This technique facilitates to achieve reduction in the volume of equipment, consumption of energy, and decreasing waste that could result in safer, cheaper, and sustainable technologies. Reactive separation is a process that takes place of combination of reaction and separation in a single equipment unit (multifunction reactors). The reaction and separation combinedly take place in equipment without having any interdependence among them. The primary application of integrated processes has been focusing on the conservation of resources, prevention of pollution, and energy management. Process integration can be categorized into mass integration and energy integration [1].

Process integration and optimization are part of process systems engineering (PSE). Advanced process integration and optimization methods have been developed in the recent past. The integration has broad applications in chemical and other allied industries. Those applications in industries require best operating conditions that minimize the investment, environmental impact, operating cost, destruction of exergy.

Reactive separation processes are integrated processes that have a combination of reaction and separation into a single unit for simultaneous production and removal of products. It involves energy integration, mass integration, and optimization, which involves enhanced reaction rate, reducing energy, and reducing solvent need. Reactive distillation, reactive absorption, reactive adsorption, and reactive extraction are the well-known techniques involved in the reactive separation process. Owing to strong interactions of chemical reactions, heat, and mass transfer, the reactive separation process behaves in a complicated fashion. The reactive separation process is encouraging process integration technologies (PI) [2].

1.1 The need for operational control of reactive separation processes

The regular process operation and control of the reactive separation process are coupled with the regulation of variables. Every reactive separation process requires a tight control system to achieve product specifications despite a lack of information on design, material property changes, equipment degradation, and external disturbances. The deviation in process variables is compensated by deviations of actuated variables.

It is desired to isolate the disturbance at localized equipment/part of the process to avoid propagation of disturbances to other processes by using the measurement of the process and its decisive property. The integration of processes such as reactive separation demands counterbalancing effects of reaction kinetics and mass transfer. The efficiency of reactive separation process depends on the implementation of right processing parameters. When the buffer for control is removed, the deviations in process variables occur by losing the measurement information.

Variations in process variables are available for controlling the process. The case of merging of reaction and separation in a single column enables the information of external flows, temperature, and concentration along with the internal temperature. The process information and manipulated variables get reduced but the control task becomes difficult. The manipulation of variables is concerned with degrees of freedom (DOF). The constricted operation degrees of freedom are likely to have an influencing effect on other process variables and independent control loops become insufficient to handle the process. Integrated processes exhibit immoderate critical process parameters as compared to the separated process that exhibits maximum concentration or conversion within the operating region. Thus, the immoderate process parameters change the sign of process gains and manifest the design of the controller more difficult.

Even though the integrated processes have the advantages of minimal equipment and minimal energy consumption, it is confronted by process control challenges. These challenges could be overcome by the implementation of complex and better-tuned regular control structures or nonlinear model-based controllers. As shown in recent studies on reactive separation processes. The industrial case studies and experiences show that process nonlinearity makes the control problems difficult and deviations result in loss of process efficiency. Better results are obtained by implementing model predictive control as compared to conventional proportional-integral-derivative (PID) controller. The model predictive control can handle the process interactions, disturbance rejection, and set-point tracking suitable for stability and robust performance in presence of process noise [3, 4].

2. Model predictive control (MPC)

Model predictive control utilizes a process model to predict future behavior while satisfying constraints in the process. Predictive technique calculates deviations from the reference. The algorithm works on repeated optimization of mathematical model of the process. The system model is used for calculating future system behavior that determines the optimal trajectory of manipulated variable (**Figure 1**) [5].

The cost index is minimized by MPC minimizes by tracking the error between reference and output

$$x(k+1) = f(x(k).u(k))$$
$$y(k) = h(x(k))$$
$$\min_u J(x(k), u(.))$$
$$\sum_{i=N_1}^{N_2} \|(r(k+i|k) - y(k+i|k\| \quad (1)$$

MPC has a diverse range of techniques encompassing dynamic matrix control (DMC), generalized predictive control (GPC), and receding horizon control (RHC). These techniques are based on their control sequence implementations and constraint

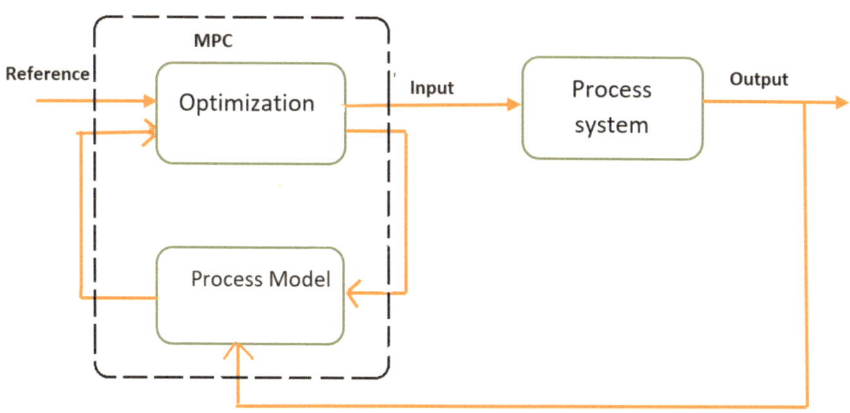

Figure 1.
A MPC-based control loop.

and model formulation. MPCs are widely implemented in Petroleum refineries and petrochemical plants.

As MPC technique is a proven control technique that can handle the state and input constraints, interaction, and measured disturbances. Modern process control techniques rely on the application of dynamic models. Intensive processes exhibit specific and dynamic characteristics. The intensified processes are directed toward optimization and model-based control [6].

This chapter explains recent advancements happened in MPC applicable to reactive separation processes and presents a case study on the implementation of Linear MPC in the natural gas sweetening process.

2.1 Significant role of MPC and its previous proven benefits in reactive separation processes

The demerits associated with conventional PID controller invoke the need for advanced controllers such as model predictive controller, non-linear model predictive controller (NLMPC). In case of reactive distillation which has a multiplicity of steady states, vapor-liquid equilibrium interaction, and a higher degree of nonlinearity, demand application of Nonlinear model predictive control and multi-parametric MPC. The employment of NMPC in batch esterification process has been considered to be suitable where ethanol and acetic acid are used as reactants to produce ethyl acetate and water. The application of nonlinear-based MPC produced stringent control of the system. The NMPC algorithm was useful for the input for providing the desired amount of distillate [7].

In case of an ideal reactive distillation column that has two ideal reactants and two products, the performance of MPC was analyzed for various scenarios. It was observed that MPC performed well in set point tracking and regulatory control. In addition to that, MPC presented a smooth variation in manipulated variables that yielded minimized energy requirement for control [8].

Optimal control based on design configuration was attained inspite of the presence of range of measured disturbances [4]. In this study, MTBE distillation column used in the study had time invariant design variables such as column diameter, catalyst distribution, number of trays, feed tray and time-variant design variables such as feed flow rate, molar flow rate, liquid methanol feed flow rate.

Application of MPC strategies for reactive distillation column used for benzene hydrogenation that was utilized in gPROMS [9]. In the study, the performance of Several MPCs was analyzed for different sets of manipulated and controlled variables analyzed by Vishal Mahindrakar and Juergen Hahn by keeping the remaining variable in feedback control. It was observed that all MPCs exhibited better responses than the PI feedback control scheme. In the case of feed benzene concentration disturbances, Single input-Single output (SISO) performed better than Multi input- Multi output (MIMO). MPC inputs with significant disturbances exhibited a reduction in an upset condition.

The implementation of ANN also quite useful in process applications. The capability of ANN to predict future behavior is quite useful to model various complicated physics-based processes. The computational cost is also less in the case of ANN model. For example in the case of a reactor where the kinetics information is not available, data-based modeling is quite useful. MPC coupled with the ANN model can estimate the behavior of nonlinear system accurately under operating variable change and disturbance. In the case of application of a neural network model predictive control

(NNMPC) was developed for depropanizer column, the NNMPC strategy predicted the output correctly which was confirmed by the low MSE index. This meant the simulation and predicted data were very small [10].

Najim and Ruiz [11] investigated the concentration of CO_2 in a gas mixture reduced to a specified value using Diethanolamine (DEA) solvent. Long-range adaptive control was achieved by using model parameters developed from the least square algorithm. The primary aim of the study was to achieve horizon-control policy by attaining minimization of quadratic criterion function composed of the input and output tracking errors in feedback. Desired performance was accomplished by the implementation of the adaptive control strategy. High sensitive absorption column efficiency was attained in this study [11].

Optimal control of amine plant using non-dominated sorting genetic algorithm-II was investigated by Behroozsarand and Shafiei [12]. In this study, a multi-objective genetic algorithm concept in concurrence with PID controller was adopted to control amine plant. The tuning of the PID controller was executed through the non-dominated sorting genetic algorithm (NSGA-II). The result showed that NSGA-II-based tuning provided optimal control of plant.

3. Control of reactive separation process: a case study with natural gas sweetening absorption process

Natural gas is one of the promising energy resources that emit fewer carbon-di-oxide (CO_2) emissions as compared to other fossil fuels such as coal and oil. The contribution of natural gas to the world energy basket is significant. The consumption of natural gas has been on positive growth. It is used as a fuel in power plants, feedstock for production of ammonia & urea, fuel in furnaces and heating applications, for heating spaces and water for cooking, non-polluting fuel, and the raw material for variety of chemical products. Various forms of NG, such as LNG, RLNG, and CNG, are used in market. The natural gas processing units deal with the production of sales specified natural gas requirement by upgrading sub-quality natural gas from reservoir.

The natural gas from the reservoir contains significant quantities of impurities such as carbon-di-oxide (CO_2), hydrogen sulfide (H_2S), mercaptans, etc. These contaminants produce various problems as follows: In the presence of water, these contaminants can form acids that corrode equipment and pipelines. The presence of CO_2 in natural gas reduces the calorific value of gas. Hence, the contaminants CO_2 and H_2S removal is essential to meet market specifications and environmental compliances. A typical consumer gas should have H_2S content of less than 4 ppmv (parts per million by volume) and CO_2 content of 1% by volume. **Figure 2** represents the scope of control applications in natural gas sweetening process.

3.1 Reactive absorption process description

The natural gas sweetening process is a reactive absorption process where a liquid solvent methyldiethanolamine (MDEA), is used to contact the gaseous phase to capture the contaminants/acid gases (H_2S and CO_2). The process is carried out at a counter-current absorber through which the gas flows upwards and liquid solvent flows downwards. Generally, the acid gases H_2S and CO_2 are captured by solvent through column. The tray in the tower provides an essential contact area between gas

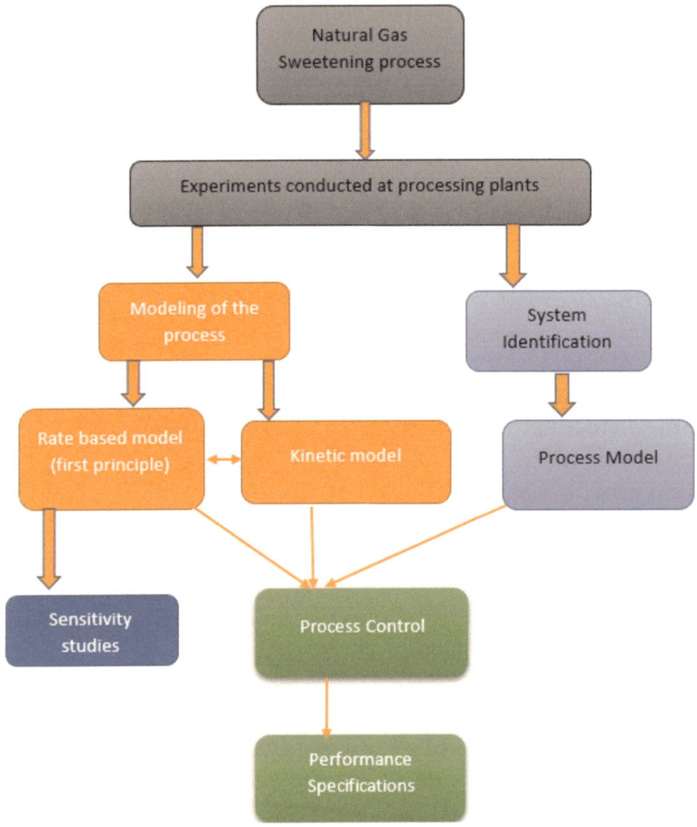

Figure 2.
Scope of process control application in natural gas sweetening.

and liquid. The residence time of reaction is governed by the flow rates of streams. Removal of any acid gases by amine washing results in the liberation of heat (heat of absorption). The diagram of acid gas absorber is given in **Figure 3** [13].

3.2 Identification of natural gas sweetening absorption process

The sensitivity studies reveals that input and output variables are involved in the amine absorption process. The transfer function of the amine absorber was developed by introducing step changes on input variables. Step responses of variables (y_1 and y_2) have been obtained using the reaction curve technique [14].

The acid gas removal efficiency can be influenced by the flow rate of solvent. The fluctuations in the flow rate of natural gas occur because of production disturbances at upstream level and acid gas content also is changed naturally depending on the reservoir conditions. The concentration of H_2S content in the outlet is influenced by MDEA flow. MDEA flow rate influences the dynamics of H_2S and CO_2 concentration in the absorber.

The typical controlled variables in the reactive absorption process are concentration and throughput of sweet gas/lean gas compositions are controlled variables. Flow rate and concentration of solvent are manipulated variables. The disturbances are feed flow rate and mole fractions in the feed gas. The control strategies for the reactive

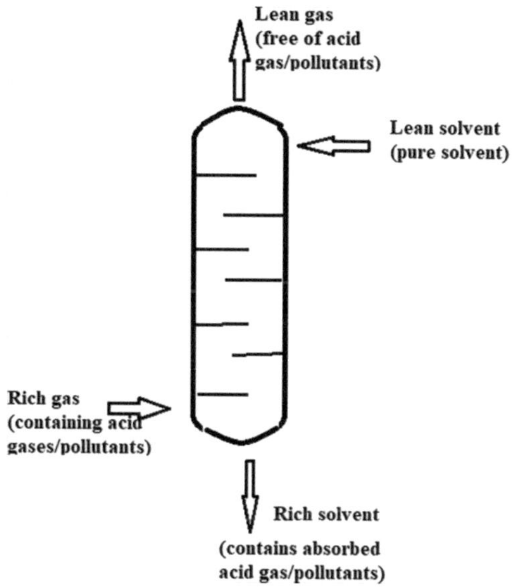

Figure 3.
A typical reactive absorption column.

absorption process have one of the following techniques: Proportional, Integral, Derivative (PID) (or)model-based (Model predictive control) (or) intelligence heuristics. The following figure presents input and output variables involved in the natural gas sweetening process (**Figures 4** and 5).

The absorber output H_2S and CO_2 content can be expressed according to (Arteaga & Contreras 1998) [15].

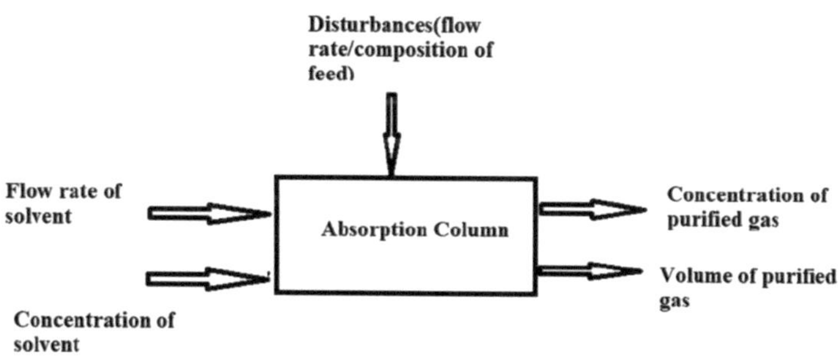

Figure 4.
Reactive absorption process input and output variables.

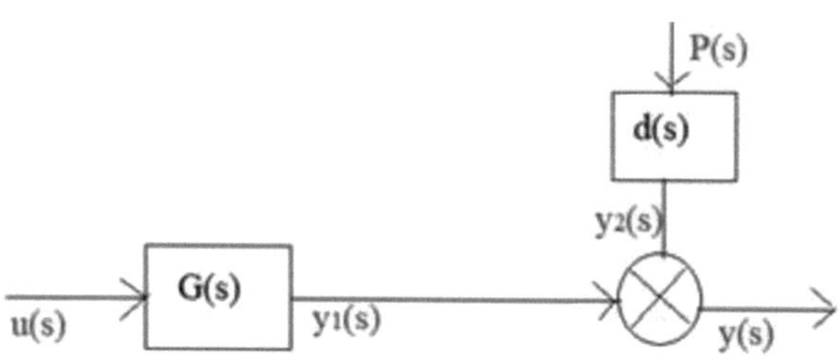

Figure 5.
Representation of amine absorber in terms of input-output variables.

$$y(s) = G(s)\Delta u(s) + d(s) + d(s)\Delta P(s) = y_1(s) + y_2(s) \quad (2)$$

u(s) = manipulated variables/input variables.
y(s) = output (H_2S and CO_2 contents in moles or ppm).
G(s) = input–output transfer function y(1).
P(s) = disturbance variable natural gas input flow.
D(s) = disturbance output $y_1(s)$ transfer function.
$Y_1(s)$ = $y_1(s)$ for P(s) = constant, $\Delta P(s) = 0$.
$Y_2(s)$ = $y_2(s)$ for U(s) = constant, $\Delta u(s) = 0$

3.3 Control of natural gas sweetening process

The control strategy for the natural gas sweetening process depends on avoiding any undesirable effects during plant operations and obtaining satisfactory performance. The frequent changes in natural gas load disturb the process as well as the feed fluctuations cause transient behavior. The control system should ensure plant to run at steady conditions to achieve desirable outputs with optimal inputs though the presence of significant uncertainty about plant behavior and disturbances. The removal of H_2S is emphasized in natural gas sweetening process and control strategy is devised based on the H_2S removal. An appropriately designed controller does ensure the solvent flow concentration and handling fluctuations in load disturbance (**Table 1**) [12].

A typical natural gas sweetening absorption column system has following controls: a flow controller performs the action of regulating the water makeup by monitoring the water amount going out of the top of the stripper; pressure controller, acts on the gas stream exiting the top of the columns; a flow controller regulates the water/solvent recycle stream from the stripper to the absorber; a level controller is employed to govern the bottom holdup of the absorber [16].

In this chapter, an IMC-PID controller and model predictive controller are used for controlling the absorption column. The performance of controllers is presented under various operating scenarios such as solvent flow rate fluctuations, solvent concentration fluctuations, as well as disturbance in gas flow rate. The dynamics of the process

S. no.	Description	Transfer function of FOPDT model
1	Transfer function relating the flow rate of solvent to concentration of acid gases (G_{11})	$\frac{-2.5957e^{-6s}}{12s+1}$
2	Transfer function relating flow rate of solvent to throughput of sweet gas (G_{12})	$\frac{-0.0053e^{-3s}}{6s+1}$
3	Transfer function relating the concentration of solvent to throughput of sweet gas (G_{22})	$\frac{0.00175e^{-3s}}{30s+1}$
4	Transfer function of the concentration of MDEA solvent to concentration of sweet gas (G_{21})	$\frac{-1.445\,e^{-6s}}{12s+1}$
5	Transfer function to flow rate of natural gas to throughput of sweet gas (d_{12})	$\frac{0.0069e^{-3s}}{6s+1}$

Table 1.
The transfer functions for amine gas sweetening process.

are studied under the inducement of a step change of ±10% to solvent flow rate, solvent concentration, and natural gas flow rate.

The manipulated variables are the MDEA(solvent) solution flow rate and MDEA solution concentration. The disturbances are the natural gas feed flow rate and the compositions of acid gases in incoming feed. The design parameters are the number of trays in the absorption column and the number of trays in the regeneration column. The H_2S removal rate, energy efficiency, and stability of plant are affected by above factors. Sweet gas compositions in the exit and throughput of sweet gas coming out of absorption tower are controlled variables.

The important variable in sweetening process is outlet H_2S concentration. The removal efficiency of H_2S is used to measure the performance of the control system. The removal efficiency is defined as the ratio of inlet to outlet H_2S concentration. A good control system guarantees the removal of H_2S by more than 90%

The control loops for the sweetening of natural gas are either PID, model-based, or intelligent heuristics. On account of their simplicity, credibility, and implementation PID controller has wide applications in industry.

Two different control strategies have been proposed for the control of natural gas sweetening columns. A 2 × 2 First Order Plus Dead Time (FOPDT) model is used for control purposes. The models are obtained through linearization around nominal operating conditions.

The PID control scheme and MPC control scheme have been investigated under flexible operating conditions concerning regulatory requirements and load changes (disturbance) in the presence of economical, environmental, and operational constraints. The control strategies performances are evaluated based on their efficiency in acid gas capturing capability, that is, the ability to reach the set point quickly without error.

3.3.1 Proportional integral derivative controller (PID) design

PID controllers are used to control the process with time delay The PID controllers are widely adopted due to their simple structure and easy implementation and maintenance. PID controller is applied when noisy signals are not observed and the dynamic response is essential. The requirement of sweetening process is to Maintain the H_2S concentration less than 2.5 moles/m^3 and limiting the absorption of CO_2 subject to the restriction of control limit. The control objective of natural gas sweetening process is to maintain the H_2S concentration actual value ($y_1(t)$) close to desired value (y^*)

The control problem of the sweetening process can be described as follows. The control action is subject to variation in the domain [$u_{1minimum}$,$u_{1maximum}$] & [$u_{2minimum}$,$u_{2\ maximum}$] & where the domain represents input manipulated variable flow rate of solvent and concentration of solvent are limited.

The general PID has the following structure

$$Gc(s) = K_c\left(1 + \frac{1}{\tau_1 s} + \tau_d s\right) \tag{3}$$

where K_c is the proportional gain, τ_I is the integral time, and τ_d is the derivative time. PID controller for absorber is designed by considering the following methodology from the literature.

The operation of the process at an unstable steady state leads to open-loop unstable systems. Due to the presence of unstable poles, the unstable process is difficult to reach stability. Because of the presence of negative or positive zeros, These systems exhibit overshoot or inverse response. The instability can be overcome by incorporating the model of the system in the controller design. This equivalent design has the robustness and it is quite promising for stabilizing higher-order systems too. A PID controller for pure integrating plus a time delay system was proposed by [17]. The controller used in the method exhibited robust performance.

The industrial time delay process transfer function model possesses one pole at origin because of their nonself regulating nature. Various methods have been proposed to tune PID controllers such as the Internal Model Control (IMC) method, direct synthesis method, empirical method, two degrees of freedom method, stability analysis method, and optimization method [18].

3.3.2 PID controller synthesis

A decentralized control system with a decoupling matrix can be designed by combining a diagonal controller $K_d(s)$ with a block compensator D(s) and the variable u_1 is manipulated by the controller in case of 2 × 2 system. This type of controller configuration provides n completely independent processes or minimal interaction processes. The purpose of decoupling is to impose calculations that cancel the existing process interactions that will provide single-loop independent control [19].

The individual loop controllers are placed diagonally in the design of decentralized control. Even though the system still exhibits interactions, the interactions are minimized. Though various types of decoupling methods are available, linear decoupling has been adopted in this work. In this, the decoupling matrices have d_{ij} components, which can eliminate interactions from all loops [19] (**Figure 6**).

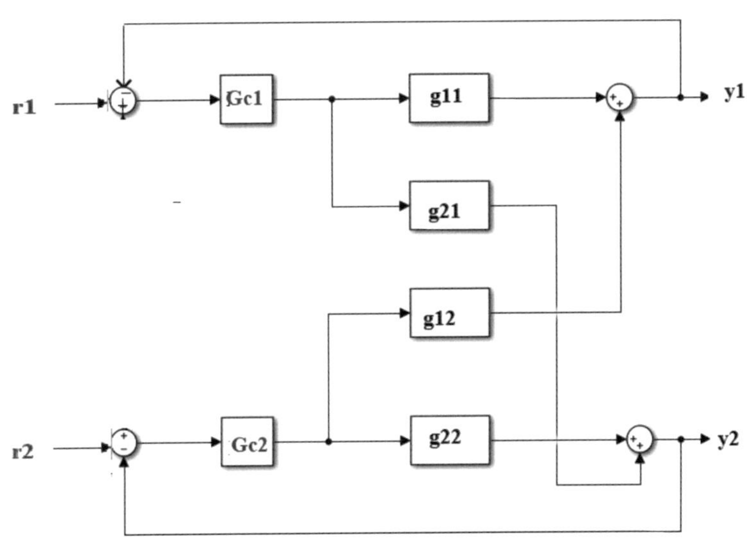

Figure 6.
The control loop diagram for the natural gas sweetening process.

The following elements are obtained for a 2 × 2 system

$$d_{11} = 1, d_{12(s)} = -\frac{G_{p12}(s)}{G_{p11}(s)}, d_{21}(s) = -\frac{G_{p12}(s)}{G_{p11}(s)}, d_{22} = 1$$

The decoupling matrix can be implemented by steady-state decoupling technique. To obtain SISO response, the transfer function of the closed loop is obtained after eliminating interactions in the decentralized structure. The transfer function of closed loop is written as follows: The controller matrix is:

$$G_c = \begin{pmatrix} G_{c11} & 0 \\ 0 & G_{c22} \end{pmatrix}$$

$$g_1 = g_{11} - G_{c12}K_{c2}(I + G_{c22}K_{c2})^{-1}G_{c21}$$

3.3.3 MPC controller synthesis for amine sweetening process

The model predictive control for amine sweetening process is represented in state space form as follows:

$$x(k+1) = \mathrm{Ax(k) + B\Delta u(k)} \quad (4)$$

$$Y(k) = \mathrm{C\,x(k) + D\,u(k)} \quad (5)$$

where

$$A = \begin{bmatrix} A & 0 \\ \mathrm{CA} & \mathrm{I} \end{bmatrix}, B = \begin{bmatrix} B_u \\ CB_u \end{bmatrix}, \; C = [0 \;\; \mathrm{I}] \text{ and } D = 0$$

x(k + 1) is the state vector at instant "k + 1"
x (k)is the state vector at instant "k"
u(k) is the input vector at instant "k"
Y(k) is the output vector at instant "k"
Δu(k) is the change in input state vector "k"
The state space matrices of gas sweetening are given by:

$$A = \begin{bmatrix} -0.08333 & 0 & 0 \\ 0 & -0.1667 & 0 \\ 0 & 0 & -0.03333 \end{bmatrix}$$

$$B = \begin{bmatrix} 0.5 & 0 \\ 0 & 0.3125 \\ 0 & 0.007813 \end{bmatrix}$$

$$C = \begin{bmatrix} 0.4328 & 0.02827 & 0 \\ 0.2408 & 0 & 0.007467 \end{bmatrix}$$

$$D = \begin{bmatrix} 0 & 0 \\ 0 & 0 \end{bmatrix}$$

Input delays (seconds): [6 3]

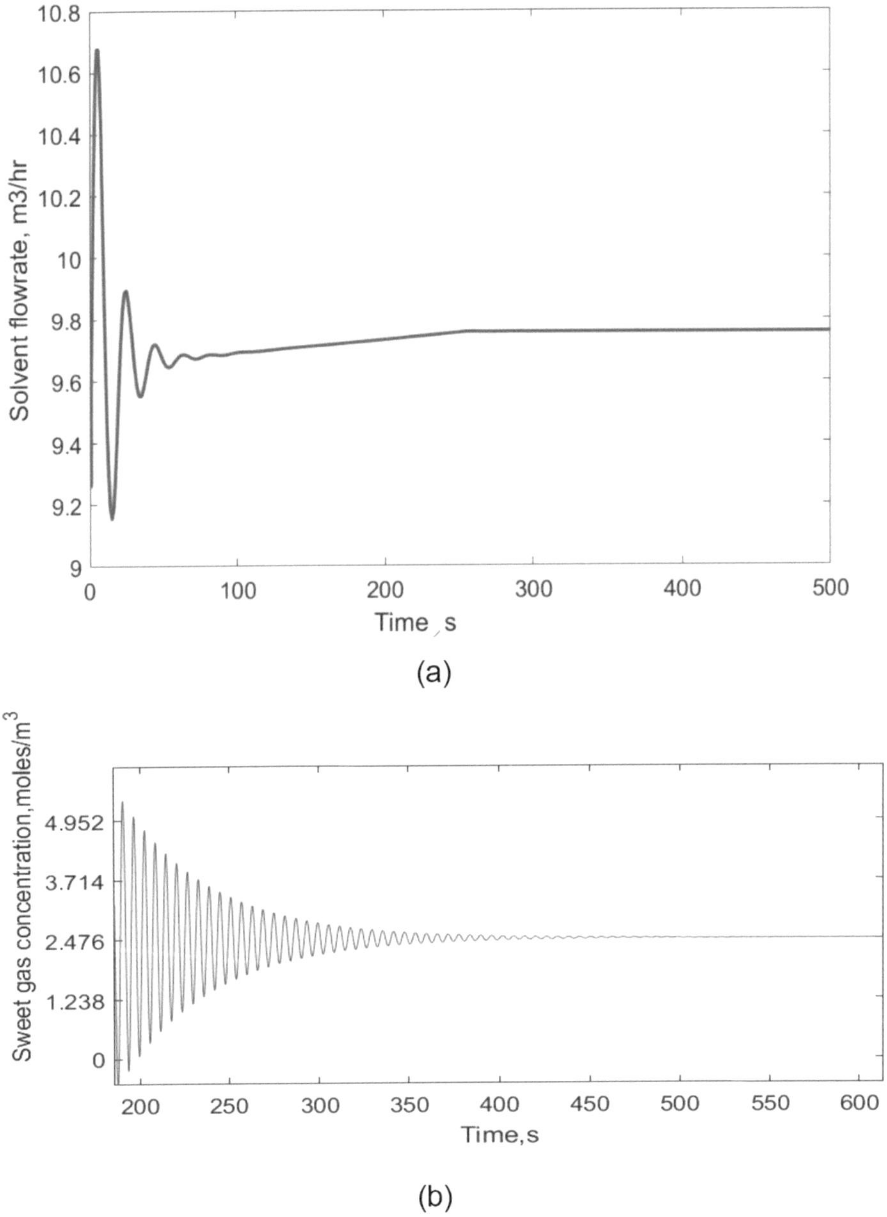

Figure 7.
(a) The solvent flow rate (manipulated variable) response and (b) the sweet gas concentration response using PID controller.

The cost function used with MPC, which penalizes the tracking error as well as the change in manipulated variable is defined in

$$J = \sum_{i=1}^{P} \| \text{r(k+1)-y(k+i)} \|^2 + \beta \sum_{i=1}^{M} \|\Delta u\|^2 \tag{6}$$

This cost function is minimized to obtain optimal values of inputs as

$$\Delta u = \left(A^T A + \beta^2 I\right)^{-1} A^T y \tag{7}$$

4. Results and discussions

The section discusses the implementation of PID and MPC control for the natural gas sweetening process and compares the performance of PID and MPC in natural gas sweetening applications.

The effects of manipulating variables such as MDEA solution flow rate and MDEA concentration variation on output variables are presented. The specification of the sweet gas concentration in the outlet gas is fixed by operational goals and must be kept within 0.5% of its set point at a steady state.

4.1 PID controller responses

A decentralized or multi-loop controller of G_{C11} and G_{C22} as components has been designed and implemented for the natural gas sweetening process. The closed loop response of acid gases (H_2S and CO_2) concentration to change or adjustment of MDEA flow rate is shown in **Figure 7a** and **b**. The closed-loop response of the concentration on account of the adjustment of MDEA concentration is shown in Figure.

4.1.1 PID controller response (y_{11})

The acid gas concentration with respect to time using PID controller was obtained and given in **Figure 6**. The response of change in concentration with respect to a longer time span is given in **Figure 7a**, to visualize the changes occur in shorter time span **Figure 7b** can be seen. It is observed from **Figure 7b**, the concentration change in acid gases is initially oscillatory. The oscillations are gradually vanished with respect to time and reaches the steady value of 2.476 moles/m^3 of acid gas after a period of 500 s.

4.1.2 PID controller response (y_{22})

In **Figure 8a** and **b**, the solvent concentration response and the controller's action on throughput of sweet gas have been given respectively. The throughput of sweet gas has evolved very slowly and takes a longer time duration to achieve steady values. The response has reached the steady value of 8.13 m^3/hr. after approximately 2.77 hours.

4.2 Model predictive control implementation

A model predictive control has been designed for natural gas sweetening absorber by considering scenarios of plant operations. The MPC controller has been assessed based on the capability of disturbance rejection and tracking about setpoint. The removal of acid gases (primarily focus on H_2S removal followed by CO_2) rate depends on the qualitative concentration of the amine solution.

The basic purpose of control strategy is to maintain acid gas concentration at 2.5 moles/m^3. The acid gas concentration is measured as output variable y_1 and throughput is measured as y_2. The following section discusses the various scenarios of plant

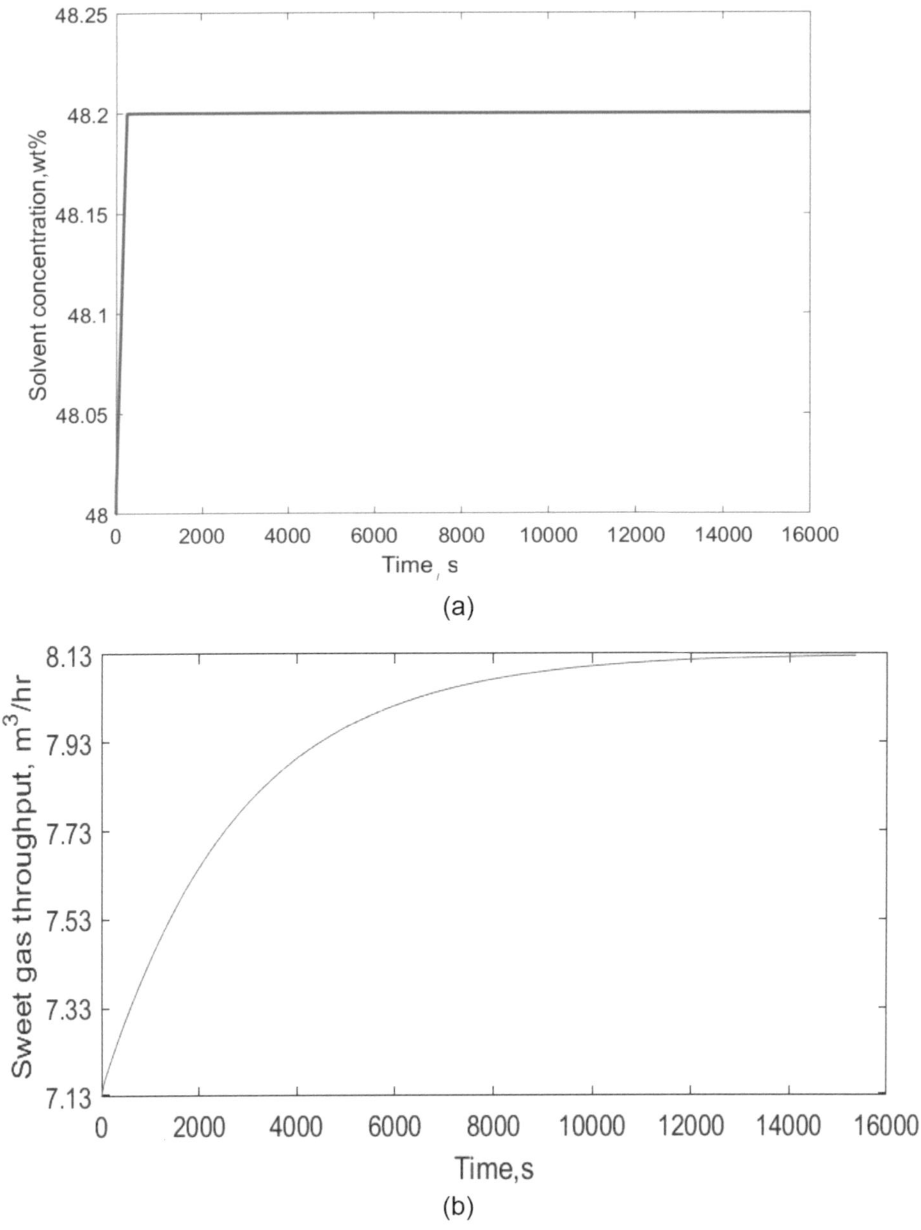

Figure 8.
(a) The solvent concentration(manipulated variable) response (b) The response of sweet gas throughput using PID controller.

operations controlled by MPC and its performance assessment in the handling of the plant under changes in input variables.

4.2.1 Setpoint tracking of the controller during closed loop response

The control objective is to maintain the output variables at the setpoint. The controlled variable is the concentration of acid gases (H_2S and CO_2). The scenario

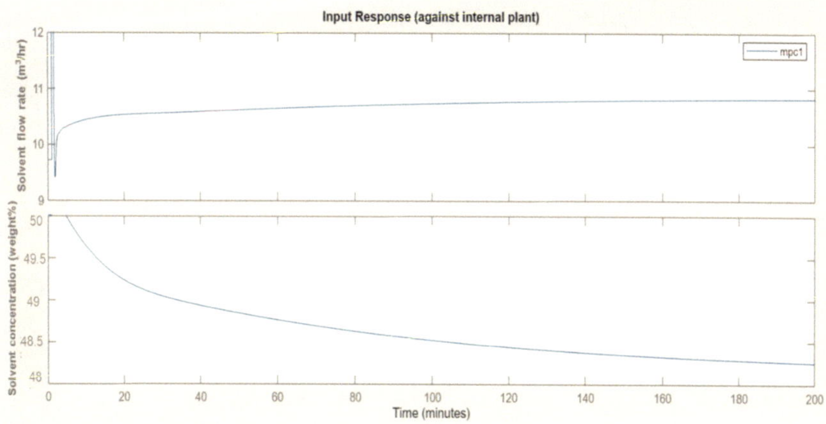

(a)

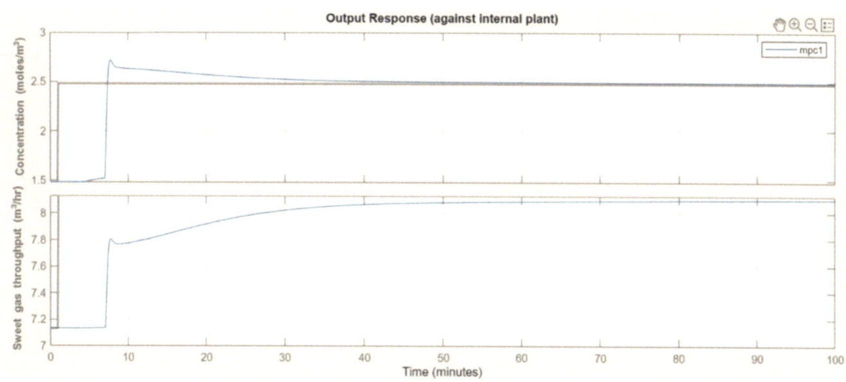

(b)

Figure 9.
(a) The movement of inputs under closed loop response (b) Set point tracking responses of sweet gas concentration (y_1) and sweet gas throughput change (y_2) under closed-loop response.

(conditions under normal plant operation) was simulated for constant solvent flow rate and constant solvent concentration.

It is observed from **Figure 9b** the MPC controller ensures the concentration of H_2S and CO_2 to reach the desired set point. In the case of throughput, sweet gas going out of absorption tower has been increased. The steady-state values have been attained after 35 minutes.

4.2.2 Set point tracking in closed loop response during step change in solvent flow rate

The following scenario has been done for manipulated step input change in solvent flow rate. The movements of manipulated variables are shown in **Figure 10**:

As we can observe from **Figure 10b** the controlled variable (sweet gas) response. This MPC simulation produced the concentration of H_2S and CO_2 to reach below its desired set point, during step change in solvent flow rate. After dead time, the concentration initially rises quickly and after 40 min, steady values have appeared. The throughput has been increased and reached a steady state value after 35 min. The change in throughput has appeared by the interaction effect with solvent flow rate.

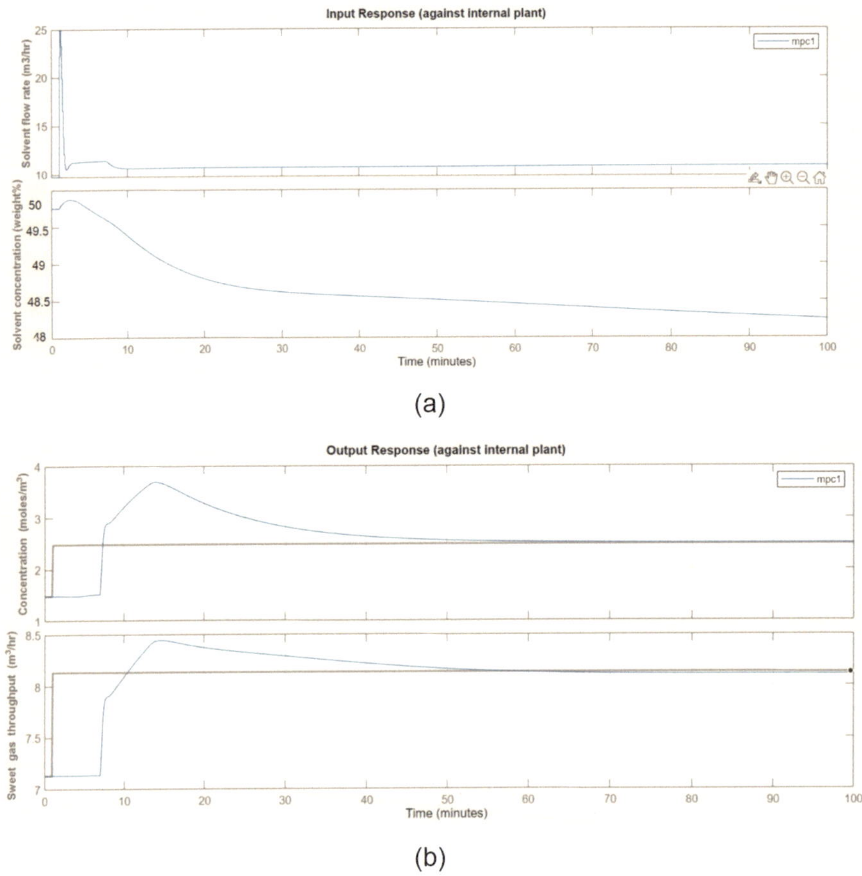

Figure 10.
(a) The input response of step change in solvent flow rate (b) Response of sweet gas concentration and sweet gas throughput due to step input change in solvent flow rate.

4.2.3 Set point tracking of concentration in the aggressive mode of controller

The following section is presented with aggressive response of MPC for the concentration control of acid gases.

We can observe from **Figure 11b** the controlled variable sweet gases concentration has stayed below 4.633 moles/m^3 for a step change in solvent flow rate. Initially, the concentration-response shot up within a shorter time duration and smoothly settled down after 40 min. The throughput also reached steady value after 35 min.

4.2.4 The response of output variables during load disturbance in closed-loop response

The scenario was analyzed for load disturbance (varying natural gas flow rate). The response of input variables is given in **Figure 11**. The solvent flow rate and concentration were adjusted according to variations in load.

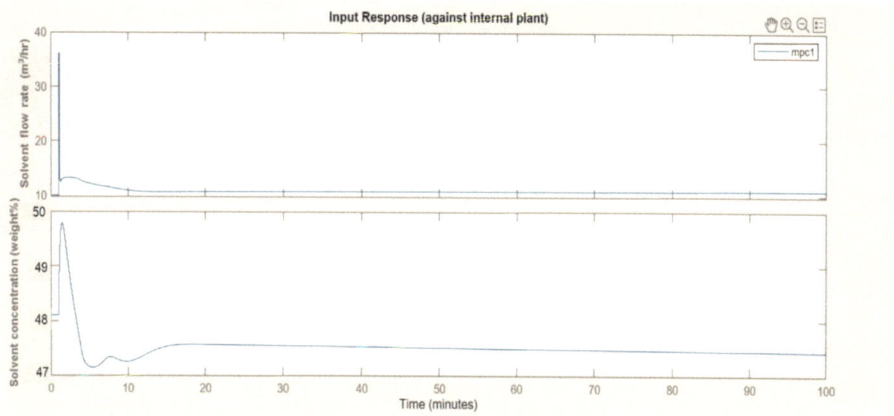

(a)

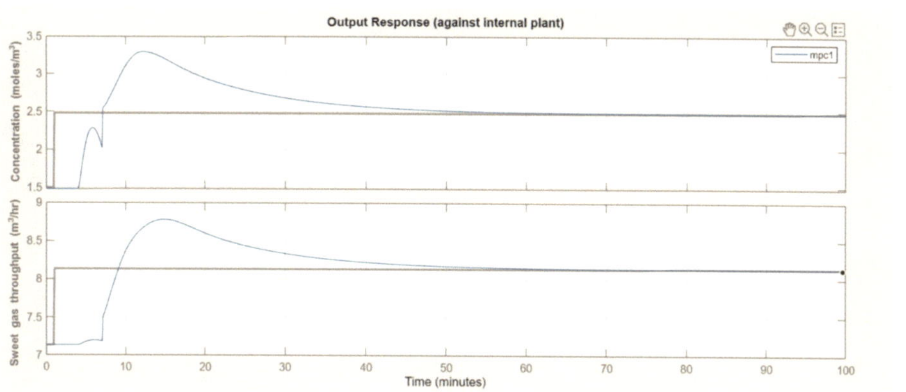

(b)

Figure 11.
(a) The movement of input variables during aggressive performance (b) The response in acid gases concentration and throughput during aggressive performance.

When there is load disturbance, the handling of output concentration and throughput is given in the **Figure 12**. The concentration stayed at 2.5 moles/m^3 and the throughput stayed at 7.15 m^3/h. It can be observed from Figure that the MPC handles the load disturbance effectively and ensures the targeted set point in both output variables. Despite variation in load, the output response shows initial overshoots after longer duration.

4.2.5 The response of output variables in aggressive mode during load disturbance

The response of output during load disturbance is given in **Figure 13**. The acid gas concentration has reached the desired specification. The set point has reached by 25 min. The concentration steady values have taken place for 25 min. The throughput settling time is less than the concentration settling time as compared to settling of concentration. When load disturbance occurred, the sweet gas going out of the absorption tower increased marginally.

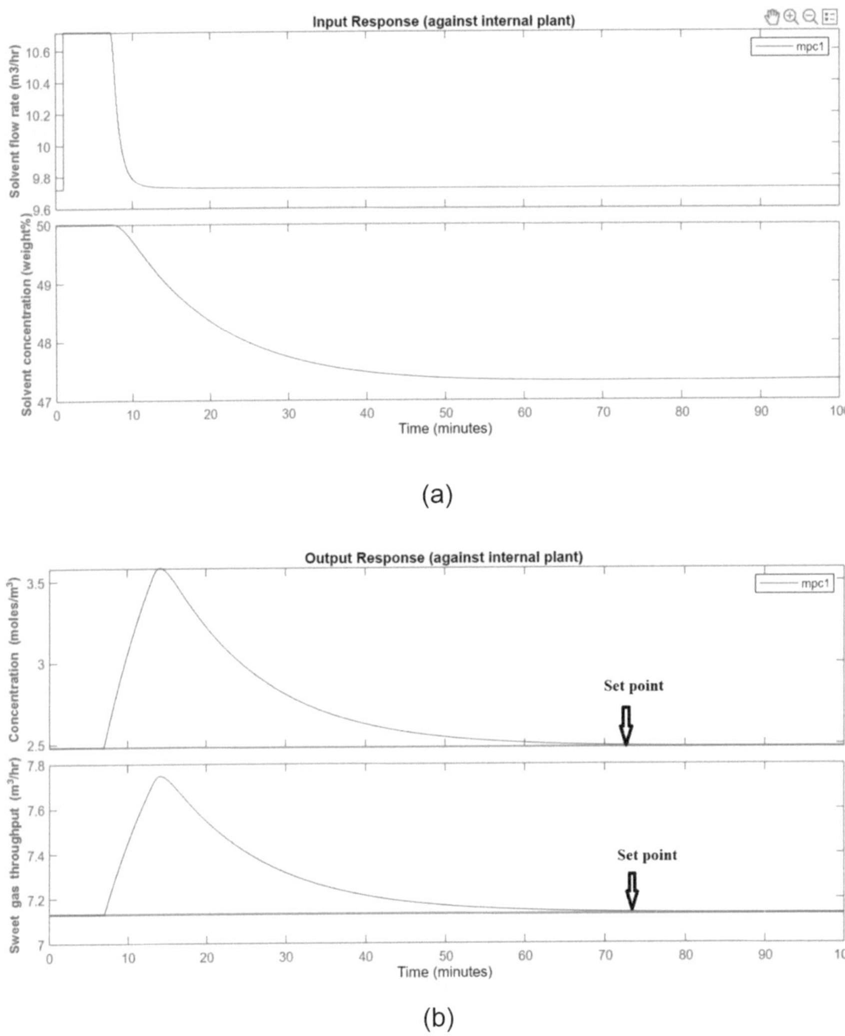

Figure 12.
(a) The input movements by the controller during load disturbances (b) The output response by controller during load disturbance.

4.3 Performance assessment of controllers

The performance of controllers can be measured by set point tracking and disturbance rejection. The performance of MPC and PID have been assessed by Integral Absolute Error (IAE) criteria which measures the integral of absolute value of deviation from setpoints. The IAE values are tabulated in **Table 2.**

The major objective of the chapter is to understand the controller's handling ability in the efficient removal of sour gases during changes in manipulated variables and load disturbances. The closed-loop response were studied using decentralized PID controller and closed-loop response, robust response, and aggressive response were analyzed by MPC. The IAE values of MPC aggressive in **Table 2** reveal that MPC aggressive mode efficiently handles the changes and has less error in maintaining the set point of concentration of sweet gas and throughput of sweet gas

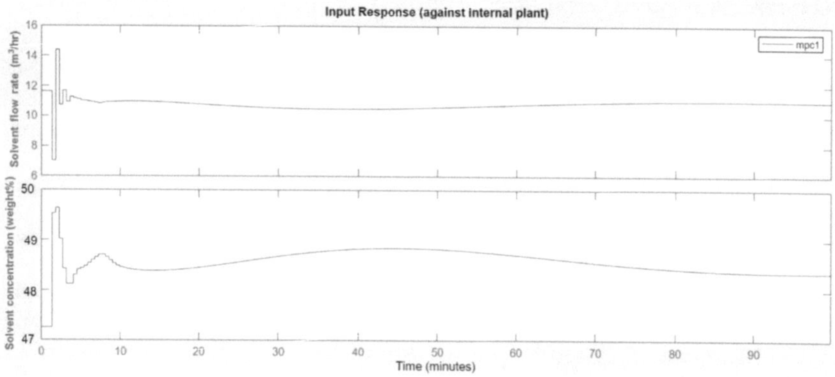

(a)

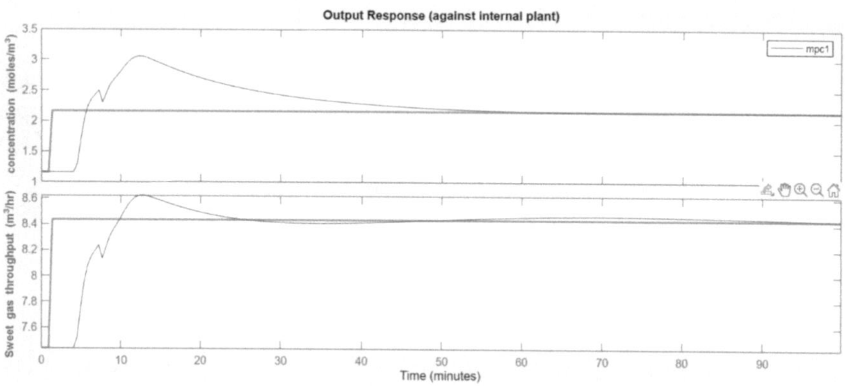

(b)

Figure 13.
(a) The input movements by the controller during aggressive mode (b) the response of output concentration and throughput during aggressive mode.

Type of controller	IAE values	
	Concentration of sweet gas	Throughput of sweet gas
PID closed loop	3.49	3.23
MPC closed loop	1.69	1.47
MPC aggressive	0.91	0.80

Table 2.
IAE values of controllers.

4.3.1 PID controller response on sweet gas concentration

The PID controller responses revealed that the acid gas concentration showed an oscillatory behavior during the initial duration and the oscillation decayed quickly. The time to reach set point is very less in the case of PID but from the IAE value, it is observed that the error is higher.

4.3.2 PID controller response on throughput of sweet gas

The PID controller responses revealed that the acid gas throughput of acid gas evolved very slowly and took a long time to reach steady state. When the concentration of solvent is reduced, the throughput of acid gas increases. The IAE value of throughput is higher than MPC.

4.3.3 MPC controller closed loop response on sweet gas concentration and throughput

The MPC controller responses revealed that the acid gas concentration reached the set point without much error. Even though the attaining of steady values has longer duration, the IAE value in closed loop response is one among the less values as compared to other MPC response IAE values. In case of throughput, the closed loop response is effective and the IAE value is less.

4.3.4 MPC controller aggressive response on sweet gas concentration and throughput

The MPC controller responses revealed that the acid gas concentration reached the set point very quickly without much error. Even though the steady values took place very faster, the IAE value in closed loop response is less comparatively than PID and MPC response. In the case of throughput, the aggressive response is very effective. The aggressive response is very much faster and more efficient than all responses.

5. Conclusions

MPC can handle multi-input multi-output (MIMO) systems that have interactions. Because of these interactions, it is often challenging to design MIMO systems using traditional controller PID. Nevertheless, MPC concurrently controls all the outputs while taking into account input-output interactions. As far as reactive absorption is concerned, MPC has been proven for tracking setpoints even though the presence of disturbances along with handling constraints. MPC satisfies process intensification objects along with energy, mass, and environmental constraints. The MPC applications in post-carbon capture have proved that Integral square error in case of step change in flue gas flow rate of MPC is less than PI controller whereas the percentage carbon capture setpoint tracking is lower in case of MPC. A huge amount of heat duty has been saved by implementing an MPC controller that PI controller. The reboiler heat duty is significantly less than the PI controller. In case of settling time for reaching steady-state values, The MPC has taken less time to reach the setpoint quickly [20, 21].

The application of MPC in post-carbon capture (PCC) has been promising. In addition to that, MPC approach has exhibited the best performance compared to conventional PID control in the natural gas sweetening process. The MPC had performed well even under disturbance by avoiding large overshoots in case of step increment in solvent flow rate and acid gas composition. MPC allows the lean solvent flow rate to attain new operating points faster than the PID controller with no fluctuations. MPC has less computational speed when it has to calculate multiple trains. MPC has dynamic adaptability in maintaining different trains to achieve different individual acid gas rates to reach the overall acid gas rate. MPC has less integral square error for overall acid gases capture control on account of step increase/decrease in the manipulated variable as well as load changes. PID scheme has sluggish performance as compared to MPC controller. PID controller took a longer time and the responses are oscillatory to reach the desired set point. Whereas the MPC controller can handle the changes in manipulated variables along with satisfying constraints and can reach a set point within shorter time horizon. In all different cases of the MPC scheme, the trade-off observed is H_2S concentration reaches nil value and concentration reduction of CO_2 is compromised even though it is reduced to some extent.

Acknowledgements

To prepare a chapter, the authors gratefully acknowledge the support of A.C. Tech and CSIR-CLRI.

Conflict of interest

The authors declare no conflict of interest.

Author details

K. Karthigaiselvan[1*] and Rames Chandra Panda[2]

1 Department of Chemical Engineering, A C Tech, Anna University, Chennai, India

2 Department of Chemical Engineering, CSIR-CLRI, Chennai, India

*Address all correspondence to: k.karthigaiselvan@gmail.com

References

[1] Florez-Orregoa D, Shivom S, Seyed N. Editorial of the research topic: Integration and optimization in the chemical process industry. Frontiers in Chemical Engineering. 82

[2] Nikačević NM, Huesman AEM, Van den Hof PMJ, Stankiewicz AI. Opportunities and challenges for process control in process intensification. Chemical Engineering and Processing: Process Intensification. 2012;**52**:1-15

[3] Engell S, Fernholz G. Control of a reactive separation process. Chemical Engineering and Processing: Process Intensification. 2003;**42**(3): 201-210

[4] Tian Y, Pappas I, Burnak B, Katz J, Pistikopoulos EN. Simultaneous design & control of a reactive distillation system–a parametric optimization & control approach. Chemical Engineering Science. 2021;**230**:116232

[5] Seborg DE, Edgar TF, Mellichamp DA, Doyle FJ III. Process Dynamics and Control. United States of America: John Wiley & Sons; 2016

[6] Eaton JW, Rawlings JB. Model-predictive control of chemical processes. Chemical Engineering Science. 1992; **47**(4):705-720

[7] Balasubramhanya LS, Doyle FJ III. Nonlinear model-based control of a batch reactive distillation column. Journal of Process Control. 2000; **10**(2–3):209-218

[8] Seban L, Kirubakaran V, Roy BK, Radhakrishnan TK. GOBF-ARMA based model predictive control for an ideal reactive distillation column. Ecotoxicology and Environmental Safety. 2015;**2015**:110-115

[9] Mahindrakar V, Hahn J. Model predictive control of reactive distillation for benzene hydrogenation. Control Engineering Practice. 2016;**52**:103-113

[10] Shin Y, Smith R, Hwang S. Development of model predictive control system using an artificial neural network: A case study with a distillation column. Journal of Cleaner Production. 2020;**277**:124124

[11] Najim K, Ruiz V. Long-range predictive control of an absorption packed column. Applied Mathematical Modelling. 1995;**19**(1):39-45

[12] Behroozsarand A, Shafiei S. Optimal control of amine plant using non-dominated sorting genetic algorithm-II. Journal of Natural Gas Science and Engineering. 2010;**2**(6):284-292

[13] Karthigaiselvan K, Panda RC. Dynamic modeling and solubility studies of sour gases during sweetening process of natural gas. Journal of Natural Gas Science and Engineering. 2021;**95**: 104087

[14] Rake H. Step response and frequency response methods. IFAC Proceedings Volumes. 1979;**12**(8):519-526

[15] Arteaga FJ, Contreras JR. System Identification and model predictive control for the optimization of a gas sweetening process. In: IEEE International Symposium on Industrial Electronics. Proceedings. ISIE'98 (Cat. No. 98TH8357). IEEE; 1998

[16] Sahraei MH, Ricardez-Sandoval LA. Simultaneous design and control of the MEA absorption process of a CO2 capture plant. Energy Procedia. Elsevier. 2014;**63**:1601-1607

[17] Panda RC. Synthesis of PID controller for unstable and integrating processes. Chemical Engineering Science. 2009;**64**(12):2807-2816

[18] Rao A, Seshagiri VSR, Rao, and M. Chidambaram. Direct synthesis-based controller design for integrating processes with time delay. Journal of the Franklin Institute. 2009;**346**(1):38-56

[19] Deshpande PB, editor. Multivariable Process Control. USA: Instrument Society of America; 1989

[20] Patron GD, Ricardez-Sandoval L. A robust nonlinear model predictive controller for a post-combustion CO2 capture absorber unit. Fuel. Elsevier. 2020;**265**:116932

[21] Zhang Q, Turton R, Bhattacharyya D. Development of model and model-predictive control of an MEA-based postcombustion CO_2 capture process. Industrial & Engineering Chemistry Research. 2016; **55**(5):1292-1308